McGRAW-HILL PUBLICATIONS IN AERONAUTICAL SCIENCE

Jerome C. Hunsaker, *Consulting Editor*

FLUID DYNAMICS

McGRAW-HILL PUBLICATIONS IN AERONAUTICAL SCIENCE

Jerome C. Hunsaker, *Consulting Editor*

FLUID DYNAMICS

By VICTOR L. STREETER

Professor of Hydraulics
University of Michigan

NEW YORK   TORONTO   LONDON

McGRAW-HILL BOOK COMPANY, INC.

1948

FLUID DYNAMICS

VII

62179

PREFACE

This book is planned to introduce the reader to the general theory of fluid flow. It is concerned primarily with the development of the general fluid equations and their reduction to specific problems. The reader is presumed to have had no mathematics beyond the elementary differential and integral calculus. Since the general treatment requires more advanced mathematics, it has been developed in each case immediately before its use is introduced. The book should be useful as a text for a second course in fluid mechanics in the senior year of college or the first year of graduate work. In preparing the material every effort has been made to clarify the concepts and to include those exasperating steps in derivations which are usually omitted.

For complete reference and more advanced treatment, the reader is referred to Lamb's "Hydrodynamics," 6th ed., Cambridge University Press, London, 1932, which is the undisputed authority in the subject and from which the author has drawn heavily in the preparation of this book.

The author has used this material in teaching advanced flow of fluids to his classes at the Illinois Institute of Technology. He wishes to acknowledge gratefully the aid given him by the David Taylor Model Basin, where the original work on the manuscript was started, and the help supplied by the Armour Research Foundation of the Illinois Institute of Technology in the computation of the flow patterns and the preparation of manuscript. Mrs. Lorraine Tuman Page is to be credited with the drawing of the flow nets, Mrs. Ruth Hopley Roberts with the preparation of the manuscript for publication, and Mr. P. C. Chu for his careful examination of the proofs. The author is deeply grateful for their help.

VICTOR L. STREETER

CHICAGO, ILL.
August, 1948

v

CONTENTS

CHAPTER I.   FLUID FLOW CONCEPTS

CHAPTER II.   FUNDAMENTALS OF FRICTIONLESS FLUID FLOW

CHAPTER III.   THEOREMS AND BASIC FLOW DEFINITIONS

CHAPTER IV.   THREE-DIMENSIONAL FLOW EXAMPLES

CHAPTER V.   APPLICATION OF COMPLEX VARIABLES
TO TWO-DIMENSIONAL FLUID FLOW

COMPLEX VARIABLES

CONFORMAL MAPPING

CHAPTER VI.   TWO-DIMENSIONAL FLOW EXAMPLES

SIMPLE CONFORMAL TRANSFORMATIONS

INVERSE TRANSFORMATIONS

CHAPTER VII. BLASIUS THEOREM—FLOW AROUND CYLINDERS AND
AIRFOILS

CHAPTER VIII. SCHWARZ-CHRISTOFFEL THEOREM—
FREE STREAMLINES

CHAPTER IX. VORTEX MOTION

THREE-DIMENSIONAL RELATIONSHIPS

TWO-DIMENSIONAL RELATIONSHIPS—RECTILINEAR VORTICES

CHAPTER X. EQUATIONS FOR VISCOUS FLOW

ANALYTICAL STATICS OF A THREE-DIMENSIONAL CONTINUUM

EQUATIONS OF MOTION

CHAPTER XI. EXAMPLES OF VISCOUS FLOW

CHAPTER XII. THE BOUNDARY LAYER

CHAPTER I

FLUID FLOW CONCEPTS

1. Flow of Matter. A fluid may be defined as a substance that deforms when subjected to a shear stress, no matter how small that shear stress may be. Fluids may be classified as Newtonian or non-Newtonian. A Newtonian fluid has a linear relation between the magnitude of applied shear stress and the rate of angular deformation of fluid. A non-Newtonian fluid has a nonlinear relation between the magnitude of applied shear stress and the rate of angular deformation.

The concepts of rate of angular deformation and magnitude of applied shear stress may be illustrated by considering the fluid between two closely spaced parallel plates so large that the conditions at the edges need not be considered. Both plates will be assumed weightless and the lower plate fixed, as shown in Fig. 1. The force F divided by the area

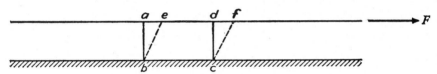

Fig. 1.—Fluid subjected to a constant shear stress.

of the upper plate is the applied shear stress τ, which may easily be seen to be constant throughout the fluid. It has been quite definitely established that the fluid immediately in contact with a solid has no motion relative to the solid.[1] When the application of a force F, no matter how small, causes the upper plate to be set in motion (with some velocity v to the right), then the substance between the two plates is a fluid.

The fluid occupying the space *abcd* is deformed into the parallelogram *ebcf*. The rate of angular deformation of the fluid is given by the angular velocity of the line containing the particles between a and b, or in other words the rate of change of the angle *abc*. In equation form the rate is given by $v/\overline{ab}$, where $\overline{ab}$ refers to the length from a to b. In general, for one-dimensional flow the rate of angular deformation is given by the change in velocity du which occurs in the distance dy perpendicular to the flow direction.

---

[1] S. Goldstein, "Modern Developments in Fluid Dynamics," Vol. II, pp. 676–680, Oxford University Press, New York, 1938.

For the Newtonian fluid

$$\tau = \mu \frac{du}{dy}$$

where μ is a constant, which is called the *viscosity*. This relationship is Newton's law of viscosity. As τ and $\frac{du}{dy}$ can be determined experimentally, μ is a derived quantity. When it is not constant for a substance that is set in motion by infinitesimal shear stresses, that substance

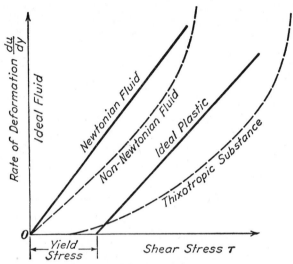

Fig. 2.—Rheological diagram.

may be classified as a non-Newtonian fluid. Many of the thick, highly viscous liquids are non-Newtonian.

The rheological diagram in Fig. 2 represents the Newtonian fluid as a straight line through the origin and the non-Newtonian fluid as a curved line through the origin. A substance may have a linear relation between shear stress and rate of deformation but sustain some finite shear stress before flow starts, as shown by the line labeled "ideal plastic." Other substances may require a finite shear stress to cause motion and may have a nonlinear relation between τ and $\frac{du}{dy}$. Many plastics are in this group. In addition, the substance may not have a fixed relation between τ and $\frac{du}{dy}$, as many substances change this relationship with the prior working or shearing that has taken place. These substances are referred to as having the property of *thixotropy*. Printers' inks are usually thixotropic.

In this work Newtonian fluids only will be considered.   Throughout most of the chapters a fictitious nonviscous fluid will be assumed in order to reduce the complexity of the mathematical treatment.   Such a fluid would be represented by the ordinate on a rheological diagram.

2. The Continuum.   Fluids are composed of molecules between which are spaces much larger than the molecules themselves.   In such a discontinuous medium the terms velocity, acceleration, density, and pressure *at a point* have no meaning.   For example, the density at a point in a fluid would be zero if that point did not happen to coincide with a molecule and would be very large if it did coincide with a molecule.   Similarly, the velocity would be zero for the first case and equal to the velocity of the molecule in the second case.

In general, average conditions at a point are required for most fluid flow cases.   This may be accomplished by replacing, for mathematical purposes, the molecular structure by a hypothetical continuous medium. In the case of fluid density at a point, defined by

$$\lim_{\Delta V \to 0} \frac{\Delta m}{\Delta V}$$

where Δm is the mass contained in the small volume ΔV, the concept of a homogeneous medium, or a continuum, is equivalent to a restriction of the meaning of the limit $\Delta V \to 0$ to ΔV approaching a small value that is still large compared with the cube of the mean free path of the fluid.

The quantities density, pressure intensity, velocity, and acceleration are assumed to vary continuously throughout the fluid except for special points, lines, or surfaces of discontinuity.

3. Stress Relationships at a Point in a Fluid.   The equations of motion for a small tetrahedron of fluid, in terms of the normal and shear stresses on its faces, yield relationships among these stresses useful in both ideal and viscous fluid flow.   The normal stress components on the faces BCO, ACO, ABO, and ABC are p_x, p_y, p_z, p, respectively, as shown in Fig. 3.   The subscript indicates that the force due to the stress is in the direction of the positive coordinate axis corresponding to the subscript.   The shear stresses τ have two subscripts, the first indicating the direction of the normal to the area over which the stress acts and the second the direction of the stress.   Let the area of the oblique face ABC be A' with l, m, n the direction cosines of its normal. p is the normal stress on A', and τ the shear stress on A'.   Let l_1, m_1, n_1 be the direction cosines of τ.

The small free body may be acted upon by two types of forces: body forces and surface forces.   The body forces act through the mass center of the tetrahedron, and their resultant may be assumed to act in the z-direction without loss of generality. Let the magnitude of the resultant body force be Z per unit mass (the most common body force is the attraction of gravity). The mass of the tetrahedron is $(\rho/6)\,\delta x\,\delta y\,\delta z$, where ρ is the mass density. Substituting into Newton's second law of motion:

Resultant force = mass times acceleration

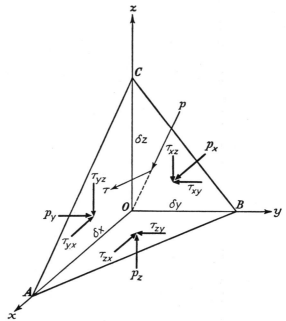

Fig. 3.—Normal and shear stresses on a tetrahedron.

written for the x-, y-, and z-components:

$$-pA'l + \tau A'l_1 + p_xA'l - \tau_{yx}A'm - \tau_{zx}A'n = \frac{\rho\,\delta x\,\delta y\,\delta z}{6}\,a_x$$

$$-pA'm + \tau A'm_1 - \tau_{xy}A'l + p_yA'm - \tau_{zy}A'n = \frac{\rho\,\delta x\,\delta y\,\delta z}{6}\,a_y$$

$$-pA'n + \tau A'n_1 - \tau_{xz}A'l - \tau_{yz}A'm + p_zA'n + \frac{Z\rho\,\delta x\,\delta y\,\delta z}{6} = \frac{\rho\,\delta x\,\delta y\,\delta z}{6}\,a_z$$

where a_x, a_y, a_z are acceleration components in the x-, y-, z-directions, respectively, and where

$$A'l = \frac{\delta y\,\delta z}{2}, \qquad A'm = \frac{\delta x\,\delta z}{2}, \qquad A'n = \frac{\delta x\,\delta y}{2}$$

Expressing the mass in terms of A', the area A' may be divided out of the equations, leaving

$$-pl + \tau l_1 + p_x l - \tau_{yx} m - \tau_{zx} n = \frac{\rho\, \delta x\, l a_x}{3}$$

$$-pm + \tau m_1 - \tau_{xy} l + p_y m - \tau_{zy} n = \frac{\rho \delta y\, m a_y}{3}$$

$$-pn + \tau n_1 - \tau_{xz} l - \tau_{yz} m + p_z n + \frac{Z\rho\, \delta z\, n}{3} = \frac{\rho\, \delta z\, n a_z}{3}$$

In order to find the stress relationships at a point, the tetrahedron is shrunk in size by allowing δx, δy, δz to approach zero, holding l, m, n constant in the limiting process.

As the mass terms are of a higher order of smallness than the stress terms, they disappear from the expressions, yielding the desired stress equations

$$\left. \begin{array}{l} -pl + \tau l_1 + p_x l - \tau_{yx} m - \tau_{zx} n = 0 \\ -pm + \tau m_1 - \tau_{xy} l + p_y m - \tau_{zy} n = 0 \\ -pn + \tau n_1 - \tau_{xz} l - \tau_{yz} m + p_z n = 0 \end{array} \right\} \qquad (1)$$

which will be useful in developing the Navier-Stokes equations for viscous fluids. Since δx, δy, δz may approach zero in any arbitrary manner, the direction cosines of the oblique face, l, m, n, may be arbitrarily chosen to specify the particular way the limit is taken. The condition that the shear stress τ is at right angles to the normal stress p may be written[1]

$$ll_1 + mm_1 + nn_1 = 0$$

Utilizing also the relation

$$l_1{}^2 + m_1{}^2 + n_1{}^2 = 1$$

five equations are obtained with which to determine the stresses on the oblique face, in terms of the five unknowns p, τ, l_1, m_1, n_1. Hence, if the shear and normal stress components are known for three mutually perpendicular planes through a point, the normal stress and shear stress can be completely determined for any plane through the point. It is shown in Sec. 108 of Chap. X that $\tau_{xy} = \tau_{yx}$, $\tau_{xz} = \tau_{zx}$, and $\tau_{yz} = \tau_{zy}$.

A nonviscous fluid has zero shear stress everywhere throughout the fluid. For this case Eqs. (1) reduce to

$$p = p_x = p_y = p_z \qquad (2)$$

showing that the pressure intensity is the same in all directions at a point. This holds regardless of the motion of the fluid.

---

[1] The angle θ between two lines having direction cosines l, m, n and l_1, m_1, n_1 is given by $\cos \theta = ll_1 + mm_1 + nn_1$.

4. Ideal Fluid. The "ideal fluid" of hydrodynamics is one that is frictionless and incompressible. A frictionless fluid has zero viscosity, *i.e.*, it cannot sustain a shear stress at any point. Hence, the fluid force acting on any elemental surface in the fluid is normal to that surface.

The ideal fluid concept greatly simplifies the mathematical treatment of flow cases. Although no such fluid actually exists, many real fluids have small viscosities, and the effects of compressibility may be small. In general, those problems dealing with large expanses of fluid will be less influenced by viscosity than those in which the fluid is narrowly confined between boundaries. For example, many conclusions concerning the motion of a solid through an ideal fluid are applicable with slight modification to the motion of an airship through the air or to the motion of a submarine through the ocean. On the other hand, ideal fluid theory gives very little information of value concerning flow through a pipe or in a narrow channel.

The assumptions of an ideal fluid are made in developing the fluid flow theory in the next chapter. They are brought in where needed to simplify the mathematical treatment. The development of viscous flow theory is undertaken in Chap. X.

Exercises

1. A substance is placed in the annular space between two concentric circular cylinders, one of which is free to rotate about its axis relative to the other. The following data were obtained for shear stress—rate of deformation:

τ, lb per ft^2	$\dfrac{du}{dy}$ per sec.
10.2	0
15.6	0
20.8	0.885
29.7	2.400
38.3	3.860
61.0	7.710

Classify the substance according to the rheological diagram. *Ans.* Ideal plastic.

2. (*a*) Find the rate of angular deformation of a fluid contained between two parallel plates 0.03 in. apart when one plate is moving relative to the other at 3.0 ft per sec.

(*b*) The shear stress on one plate is 1.44 psi. What is the viscosity in consistent English units?   *Ans.* (*a*) 1200 radians per sec; (*b*) 0.1728 lb-sec per ft^2.

3. Selecting as a free body a unit length of triangular prism (two-dimensional case), derive the expressions for p and τ on the inclined face (making an angle θ with the x axis) in terms of p_x, p_y, τ_{xy}, τ_{yx}. All stresses are parallel to the xy-plane.

4. Show that $\tau_{xy} = \tau_{yx}$ for the two-dimensional case by using the equation, summation of torques about an axis through the mass center equals the product of the moment of inertia about this axis and the angular acceleration.

CHAPTER II

FUNDAMENTALS OF FRICTIONLESS FLUID FLOW

5. Partial Derivatives and Total Differentials. Although partial derivatives and total differentials are included in the first year of calculus, they are not usually introduced until the latter part of the course; hence, little opportunity is available to the student to comprehend their value. As they are needed throughout this work, their definitions are given at this point.[1]

Consider

$$u = f(x,y)$$

where x,y are independent variables. If y is held constant, u becomes a function of x alone and its derivative may be determined as if u were a function of one variable. This is denoted by

$$\frac{\partial f}{\partial x} \quad \text{or} \quad \frac{\partial u}{\partial x}$$

and is called the partial derivative of f with respect to x or the partial derivative of u with respect to x. Similarly, if x is held constant, u becomes a function of y alone and $\frac{\partial f}{\partial y}$ is called the partial of f with respect to y. These partials are defined by

$$\frac{\partial u}{\partial x} = \frac{\partial f(x,y)}{\partial x} = \lim_{\Delta x \to 0} \frac{f(x + \Delta x, y) - f(x,y)}{\Delta x}$$

$$\frac{\partial u}{\partial y} = \frac{\partial f(x,y)}{\partial y} = \lim_{\Delta y \to 0} \frac{f(x, y + \Delta y) - f(x,y)}{\Delta y}$$

Examples:

1. $u = x^3 + x^3 y^2 + 3y$

$$\frac{\partial u}{\partial x} = 3x^2 + 3x^2 y^2$$

$$\frac{\partial u}{\partial y} = 2x^3 y + 3$$

2. $u = \sin (ax^2 + by^2)$

$$\frac{\partial u}{\partial x} = 2ax \cos (ax^2 + by^2)$$

$$\frac{\partial u}{\partial y} = 2by \cos (ax^2 + by^2)$$

---

[1] I. S. and E. S. Sokolnikoff, "Higher Mathematics for Engineers and Physicists," 2d ed., pp. 125–138, McGraw-Hill Book Company, Inc., New York, 1941.

In each case the differentiation is carried out exactly as for a function of one independent variable with the other independent variable considered as a constant.

If $u = f(x)$, the derivative of u with respect to x is defined by

$$\frac{du}{dx} = \lim_{\Delta x \to 0} \frac{\Delta u}{\Delta x} = \lim_{\Delta x \to 0} \frac{f(x + \Delta x) - f(x)}{\Delta x}$$
$$= f'(x)$$

Hence,

$$\Delta u = f'(x)\,\Delta x + \epsilon\,\Delta x$$

where ϵ is an infinitesimal that vanishes with Δx. Then

$$du = f'(x)\,\Delta x \equiv f'(x)\,dx$$

is the differential du.

If $u = f(x,y)$, the differential du is defined in a similar manner. If x and y take on increments Δx, Δy, then

$$\Delta u = f(x + \Delta x, y + \Delta y) - f(x,y)$$

Δx and Δy may approach zero in any manner. If Δu approaches zero regardless of the way in which Δx and Δy approach zero, then $u = f(x,y)$ is called a continuous function of x and y. It will be assumed that $f(x,y)$ is continuous and also that $\frac{\partial f}{\partial x}$ and $\frac{\partial f}{\partial y}$ are continuous.

Adding and subtracting $f(x, y + \Delta y)$ to the expression for Δu above,

$$\Delta u = f(x + \Delta x, y + \Delta y) - f(x, y + \Delta y) + f(x, y + \Delta y) - f(x,y)$$

Since

$$\lim_{\Delta x \to 0} \frac{f(x + \Delta x, y + \Delta y) - f(x, y + \Delta y)}{\Delta x} = \frac{\partial f(x, y + \Delta y)}{\partial x}$$

then

$$f(x + \Delta x, y + \Delta y) - f(x, y + \Delta y) = \frac{\partial f(x, y + \Delta y)}{\partial x} \Delta x + \epsilon_1\,\Delta x$$

where $\lim_{\Delta x \to 0} \epsilon_1 = 0$. Furthermore,

$$\lim_{\Delta y \to 0} \frac{\partial f(x, y + \Delta y)}{\partial x} = \frac{\partial f(x,y)}{\partial x}$$

as the derivative is continuous. Therefore,

$$\frac{\partial f(x, y + \Delta y)}{\partial x} = \frac{\partial f(x,y)}{\partial x} + \epsilon_2$$

where $\lim_{\Delta y \to 0} \epsilon_2 = 0$. Similarly,

$$f(x, y + \Delta y) - f(x,y) = \frac{\partial f(x,y)}{\partial y} \Delta y + \epsilon_3\,\Delta y$$

where $\lim\limits_{\Delta y \to 0} \epsilon_3 = 0$.

Therefore, it follows that

$$\Delta u = \frac{\partial f(x,y)}{\partial x} \Delta x + \frac{\partial f(x,y)}{\partial y} \Delta y + (\epsilon_1 + \epsilon_2)\Delta x + \epsilon_3 \, \Delta y.$$

In the limit as Δx and Δy approach zero, the expression for the total differential of u is obtained:

$$du = \frac{\partial f}{\partial x} \, dx + \frac{\partial f}{\partial y} \, dy$$

In general, if $u = f(x,y,z,t)$, then

$$du = \frac{\partial f}{\partial x} \, dx + \frac{\partial f}{\partial y} \, dy + \frac{\partial f}{\partial z} \, dz + \frac{\partial f}{\partial t} \, dt$$

which may also be written in the form

$$du = \frac{\partial u}{\partial x} \, dx + \frac{\partial u}{\partial y} \, dy + \frac{\partial u}{\partial z} \, dz + \frac{\partial u}{\partial t} \, dt$$

If, now, in $u = f(x,y)$, x and y are functions of one independent variable, say t, then u becomes a function of t alone and u may have a derivative with respect to t. Let $x = f_1(t)$, $y = f_2(t)$, where the functions are assumed differentiable. If t is given an increment, then x, y, and u will have corresponding increments Δx, Δy, and Δu, which approach zero with Δt. As before,

$$\Delta u = \frac{\partial f}{\partial x} \Delta x + \frac{\partial f}{\partial y} \Delta y + (\epsilon_1 + \epsilon_2) \, \Delta x + \epsilon_3 \, \Delta y.$$

Then

$$\frac{\Delta u}{\Delta t} = \frac{\partial f}{\partial x} \frac{\Delta x}{\Delta t} + \frac{\partial f}{\partial y} \frac{\Delta y}{\Delta t} + (\epsilon_1 + \epsilon_2) \frac{\Delta x}{\Delta t} + \epsilon_3 \frac{\Delta y}{\Delta t}$$

and

$$\frac{du}{dt} = \frac{\partial f}{\partial x} \frac{dx}{dt} + \frac{\partial f}{\partial y} \frac{dy}{dt}$$

Also,

$$du = \frac{\partial f}{\partial x} \, dx + \frac{\partial f}{\partial y} \, dy$$

holds, as before, even when x and y are functions of t. Moreover, if $u = f(x,y,z,t)$, where x,y,z are functions of t, then

$$\frac{du}{dt} = \frac{\partial u}{\partial x} \frac{dx}{dt} + \frac{\partial u}{\partial y} \frac{dy}{dt} + \frac{\partial u}{\partial z} \frac{dz}{dt} + \frac{\partial u}{\partial t}$$

may be shown in a similar manner.

Examples:

1. Let $u = 2x + 3y^2 + 4z^3 + tx$. Then

$$du = (2 + t) \, dx + 6y \, dy + 12z^2 \, dz + x \, dt$$

as

$$\frac{\partial u}{\partial x} = 2 + t, \qquad \frac{\partial u}{\partial y} = 6y, \qquad \frac{\partial u}{\partial z} = 12z^2, \qquad \frac{\partial u}{\partial t} = x$$

If x, y, and z are functions of t, then

$$\frac{du}{dt} = (2 + t) \frac{dx}{dt} + 6y \frac{dy}{dt} + 12z^2 \frac{dz}{dt} + x$$

Specifically, if

$$x = \sin t, \qquad y = \ln t, \qquad z = 3t^2$$

where ln represents the natural logarithm, then

$$\frac{dx}{dt} = \cos t, \qquad \frac{dy}{dt} = \frac{1}{t}, \qquad \frac{dz}{dt} = 6t$$

and

$$\frac{du}{dt} = (2 + t) \cos t + \frac{6}{t} \ln t + 648t^5 + \sin t$$

2. Let $u = xyzt$. Then

$$du = yzt \, dx + xzt \, dy + xyt \, dz + xyz \, dt$$

If

$$x = 2t, \qquad y = t^2, \qquad z = t^3$$

then

$$\frac{dx}{dt} = 2, \qquad \frac{dy}{dt} = 2t, \qquad \frac{dz}{dt} = 3t^2$$

and

$$\frac{du}{dt} = 2t^6 + 4t^6 + 6t^6 + 2t^6 = 14t^6.$$

6. Euler's Equations of Motion. Every particle of a fluid must obey Newton's second law at all times. Certain conditions (called *boundary conditions*) must also be fulfilled at the boundaries for specific flow cases. In addition, the continuity equation must be satisfied at all points in the fluid except so-called "singular points."

The boundary conditions usually take the form of an analytical expression which states that the fluid velocity component at a boundary normal to the boundary is everywhere equal to the velocity of the boundary normal to itself. This is a mathematical statement that the fluid cannot penetrate the boundary surface and that no void spaces or gaps may occur between fluid and boundary. No restriction is placed on the tangential component of the velocity for a nonviscous fluid.

The continuity equation is a restriction on the velocity distribution throughout a fluid. For an incompressible fluid it is a statement that the flow into any fixed space in the fluid must exactly equal the flow out of that same space in the same time interval.   Euler's dynamical equations for a nonviscous fluid are derived in this section, the equation of continuity in Sec. 7, and the boundary condition equations in Sec. 8.

Euler's equations of motion state that the resultant force on any fluid particle must always equal the product of the mass of the particle and its acceleration and that the acceleration is in the direction of the resultant force.   Rather than write the vector equation, expressions are developed for the xyz-components of that equation.

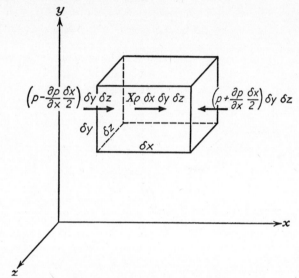

FIG. 4.—Free body showing the forces acting in the x-direction.

For convenience, consider a small parallelepiped of fluid whose edges are δx, δy, δz, respectively, parallel to the xyz-axes of a cartesian coordinate system as shown in Fig. 4.   The equations of motion are written for this small free body, and then the limiting process is applied to its edges, shrinking it to a point.   Two types of forces act on the free body: surface forces and body (or extraneous) forces.   Assuming the fluid is nonviscous, the contact forces become normal pressure forces acting on the faces of the parallelepiped.   Let the center of the body be at (x,y,z) and the pressure intensity there be p.   The equation of motion for the x-component is derived in detail.   The two other components may be obtained in a similar manner or by cyclic permutations.

As the pressure intensity is assumed to vary continuously throughout the fluid, it may take a form similar to that shown in Fig. 5, which illustrates its variation along a line parallel to the x-axis through the center of the parallelepiped. To find the approximate pressure intensity at the left face, it may be assumed that the tangent to the curve at (x,y,z) replaces the curve. The slope of the tangent is $\dfrac{\partial p}{\partial x}$ and is negative as shown in Fig. 5. Then the pressure intensity at the center of the left face is

$$p - \frac{\partial p}{\partial x}\frac{\delta x}{2}$$

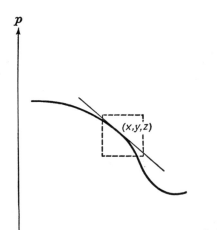

and at the right face is

$$p + \frac{\partial p}{\partial x}\frac{\delta x}{2}$$

The error in these expressions is visible in Fig. 5, caused by the divergence of the curve and its tangent. When δx is allowed to approach zero, the curve and tangent coincide and the above expressions become exact. Similar reasoning applies to variations in pressure intensities over the faces of the parallelepiped; and likewise, these expressions become exact in

Fig. 5.—Illustration of pressure variation along a line parallel to the x-axis through the center of the parallelepiped.

the limit. The resultant surface force in the positive x-direction is

then

$$\left(p - \frac{\partial p}{\partial x}\frac{\delta x}{2}\right)\delta y\ \delta z - \left(p + \frac{\partial p}{\partial x}\frac{\delta x}{2}\right)\delta y\ \delta z$$

or, simplifying,

$$-\frac{\partial p}{\partial x}\ \delta x\ \delta y\ \delta z$$

Let the component of the body forces per unit mass in the positive x-direction be X. The extraneous force in the x-direction acting on the free body is

$$X\rho\ \delta x\ \delta y\ \delta z$$

where ρ is the mass density of the fluid. Adding the surface and body force components together and equating to the mass times the acceleration in the x-direction yield

$$\left(X\rho - \frac{\partial p}{\partial x}\right)\delta x\ \delta y\ \delta z = \rho\ \delta x\ \delta y\ \delta z\ a_x$$

where a_x is the acceleration component.

Now, allow δx, δy, δz to approach zero and divide through by the mass of the parallelepiped, yielding

$$X - \frac{1}{\rho}\frac{\partial p}{\partial x} = a_x \tag{1a}$$

In an analogous manner, the y and z equations are obtained

$$Y - \frac{1}{\rho}\frac{\partial p}{\partial y} = a_y \tag{1b}$$

$$Z - \frac{1}{\rho}\frac{\partial p}{\partial z} = a_z \tag{1c}$$

where Y, Z are the components of the body forces per unit mass in the y- and z-directions, respectively, and a_y, a_z are the corresponding acceleration components.

In order to obtain expressions for a_x, a_y, and a_z, general functional relations of the velocity components are used. Let u, v, w be the velocity components in the xyz-directions at the point x,y,z. In general, u, v, and w are functions of the coordinates and of the time t; thus

$$u = u(x,y,z,t)$$
$$v = v(x,y,z,t)$$
$$w = w(x,y,z,t)$$

where x, y, z, and t are independent variables. If, however, the displacement, velocity, and acceleration of the particle at (x,y,z) at time t is desired, then x, y, z become functions of the time t. Writing the total differential for u,

$$du = \frac{\partial u}{\partial x}\,dx + \frac{\partial u}{\partial y}\,dy + \frac{\partial u}{\partial z}\,dz + \frac{\partial u}{\partial t}\,dt$$

from Sec. 5. Dividing by dt,

$$a_x = \frac{du}{dt} = \frac{\partial u}{\partial x}\frac{dx}{dt} + \frac{\partial u}{\partial y}\frac{dy}{dt} + \frac{\partial u}{\partial z}\frac{dz}{dt} + \frac{\partial u}{\partial t}$$

As x, y, z represent the coordinates of the particle as it moves, $\dfrac{dx}{dt}$, $\dfrac{dy}{dt}$, and $\dfrac{dz}{dt}$ are the velocity components u, v, and w, respectively. Hence

$$a_x = u\frac{\partial u}{\partial x} + v\frac{\partial u}{\partial y} + w\frac{\partial u}{\partial z} + \frac{\partial u}{\partial t} \tag{2a}$$

and

$$a_y = u\frac{\partial v}{\partial x} + v\frac{\partial v}{\partial y} + w\frac{\partial v}{\partial z} + \frac{\partial v}{\partial t} \tag{2b}$$

$$a_z = u\frac{\partial w}{\partial x} + v\frac{\partial w}{\partial y} + w\frac{\partial w}{\partial z} + \frac{\partial w}{\partial t} \tag{2c}$$

Substituting these expressions for the accelerations in Eqs. (1), the Euler equations are obtained:

$$
\left.
\begin{aligned}
X - \frac{1}{\rho}\frac{\partial p}{\partial x} &= u\frac{\partial u}{\partial x} + v\frac{\partial u}{\partial y} + w\frac{\partial u}{\partial z} + \frac{\partial u}{\partial t} \\
Y - \frac{1}{\rho}\frac{\partial p}{\partial y} &= u\frac{\partial v}{\partial x} + v\frac{\partial v}{\partial y} + w\frac{\partial v}{\partial z} + \frac{\partial v}{\partial t} \\
Z - \frac{1}{\rho}\frac{\partial p}{\partial z} &= u\frac{\partial w}{\partial x} + v\frac{\partial w}{\partial y} + w\frac{\partial w}{\partial z} + \frac{\partial w}{\partial t}
\end{aligned}
\right\}
\qquad (3)
$$

No assumption has been made that ρ is a constant; hence, these equations apply equally well to compressible and incompressible non-viscous fluids.

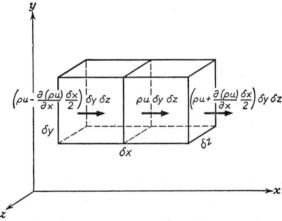

FIG. 6.—Mass flow through the faces of a parallelepiped that are normal to the x-axis

7. Equation of Continuity. In deriving the equation of continuity it is convenient to consider the mass flow through the faces of a small parallelepiped with edges parallel to an xyz-coordinate system and center at (x,y,z). The center is fixed in space. Referring to Fig. 6, the expression for rate of increase in mass of fluid within the parallelepiped with edges δx, δy, δz is formulated. As no restrictive assumptions are made, the resulting equation applies to any fluid, real or ideal.

Fixing attention first on the rate of increase in mass due to flow through the two faces normal to the x-axis, the mass of fluid per unit time passing the face through the center point (x,y,z) and normal to the x-axis is given approximately by

$$\rho u \, \delta y \, \delta z$$

where again ρ is the mass per unit volume at (x,y,z) and u is the x-component of the velocity at the center. It is an approximate expression because ρu for the center point may not be the mean value over the finite area $\delta y\ \delta z$. The mass flow per unit time through the face nearer the origin is given approximately by

$$\left[\rho u - \frac{\partial(\rho u)}{\partial x}\frac{\delta x}{2}\right]\delta y\ \delta z$$

and the mass flow per unit time *into* the face farther from the origin is given approximately by

$$-\left[\rho u + \frac{\partial(\rho u)}{\partial x}\frac{\delta x}{2}\right]\delta y\ \delta z$$

The minus sign is required, since u is taken positive in the positive x-direction. The approximate net mass inflow per unit time to the parallelepiped through the two faces is the algebraic sum of the last two expressions, or

$$-\frac{\partial(\rho u)}{\partial x}\delta x\ \delta y\ \delta z$$

In an analogous manner the rates of mass inflow through the other two pairs of faces normal to the y- and z-axes, respectively, are given by

$$-\frac{\partial(\rho v)}{\partial y}\delta x\ \delta y\ \delta z \qquad \text{and} \qquad -\frac{\partial(\rho w)}{\partial z}\delta x\ \delta y\ \delta z$$

where v, w are y- and z-components of velocity at (x,y,z). Adding the three rates together, the approximate net mass inflow per unit time is

$$-\left[\frac{\partial(\rho u)}{\partial x} + \frac{\partial(\rho v)}{\partial y} + \frac{\partial(\rho w)}{\partial z}\right]\delta x\ \delta y\ \delta z$$

The net increase in mass per unit time in the parallelepiped is given approximately by

$$\frac{\partial}{\partial t}(\rho\ \delta x\ \delta y\ \delta z)$$

As δx, δy, δz are independent of the time, this may be written

$$\frac{\partial\rho}{\partial t}\delta x\ \delta y\ \delta z$$

Each of the two expressions for increase in mass per unit time are approximate because of the possible nonlinear space rate of change of ρ, u, v, w, in a manner analogous to the pressure change illustrated in Fig. 5. If δx, δy, δz arc allowed to become very small, then the expressions become more accurate and are exact in the limit as the parallelepiped is shrunk to the point (x,y,z). Applying the limiting process and dividing through by the mass of the parallelepiped, the general continuity equation in cartesian coordinates is obtained:

$$-\frac{1}{\rho}\left[\frac{\partial(\rho u)}{\partial x} + \frac{\partial(\rho v)}{\partial y} + \frac{\partial(\rho w)}{\partial z}\right] = \frac{1}{\rho}\frac{\partial\rho}{\partial t} \qquad (4)$$

This equation must necessarily hold for every point in the fluid with the exception of the so-called "singular points" such as sources and sinks, which are discussed in Chap. III.

If the density is constant, the right-hand side of Eq. (4) vanishes. Equation (4) reduces to the kinematical expression

$$\frac{\partial u}{\partial x} + \frac{\partial v}{\partial y} + \frac{\partial w}{\partial z} = 0 \qquad (5)$$

which is the equation of continuity for an ideal fluid.

The ideal fluid assumptions have led to four equations that must be satisfied at every point in the fluid (except singular points). These partial differential equations, Eqs. (3) and (5), contain four unknowns u, v, w, and p, all of which are functions of the coordinates and the time. As arbitrary functions enter the solutions of partial differential equations, boundary conditions are also required for specific solutions of flow cases.

Considering a frictionless fluid of variable density, Eqs. (3) and (4) are available in five unknowns: u, v, w, p, and ρ. The density is usually expressed in terms of the pressure to obtain an additional equation, called the *equation of state* of the fluid. Examples of pressure-density relationships are those for isothermal or adiabatic expansion or compression.

8. Boundary Conditions. The kinematical condition that must be satisfied at every point on a solid-fluid boundary is that the fluid does not penetrate the boundary and that no gaps occur between boundary and fluid. If the boundary is *stationary*, this may be stated as

$$lu + mv + nw = 0 \qquad (6)$$

or that the component of the fluid velocity normal to the surface must be zero. l, m, n are direction cosines of the normal to the surface, and u, v, w the velocity components used heretofore.

When the boundary *itself* is in motion, however, then the fluid velocity component normal to the boundary must equal the velocity of the boundary normal to itself. Letting $\dot{\nu}$ (the derivative of ν with respect to t) represent this velocity of the boundary normal to itself, the boundary condition becomes

$$\dot{\nu} = lu + mv + nw \qquad (7)$$

for every point on the boundary.

The equation of the boundary surface may be expressed as

$$F(x,y,z,t) = 0 \tag{8}$$

To find the rate of motion normal to itself at any point (x,y,z) on this surface, first write the total derivative of Eq. (8)

$$dF = \frac{\partial F}{\partial x}\, dx + \frac{\partial F}{\partial y}\, dy + \frac{\partial F}{\partial z}\, dz + \frac{\partial F}{\partial t}\, dt = 0 \tag{9}$$

To follow the motion normal to the surface, let

$$dx = l\dot{\nu}\, \delta t, \qquad dy = m\dot{\nu}\, \delta t, \qquad dz = n\dot{\nu}\, \delta t, \qquad dt = \delta t \tag{10}$$

Substituting into Eq. (9) and reducing,

$$\dot{\nu}\left(l\frac{\partial F}{\partial x} + m\frac{\partial F}{\partial y} + n\frac{\partial F}{\partial z}\right) + \frac{\partial F}{\partial t} = 0 \tag{11}$$

As the direction cosines of the normal to the surface [1] at (x,y,z) are

$$l = \frac{1}{R}\frac{\partial F}{\partial x}, \qquad m = \frac{1}{R}\frac{\partial F}{\partial y}, \qquad n = \frac{1}{R}\frac{\partial F}{\partial z} \tag{12}$$

where

$$R = \sqrt{\left(\frac{\partial F}{\partial x}\right)^2 + \left(\frac{\partial F}{\partial y}\right)^2 + \left(\frac{\partial F}{\partial z}\right)^2}$$

[1] Let $f(x,y,z) = 0$ represent the boundary surface $F(x,y,z,t) = 0$ at some time t_0. The expressions for the direction cosines of the normal to the surface at some point (x_0,y_0,z_0) are developed as follows:

Taking the total derivative of $f(x,y,z) = 0$,

$$\frac{\partial f}{\partial x}\, dx + \frac{\partial f}{\partial y}\, dy + \frac{\partial f}{\partial z}\, dz = 0$$

Consider two neighboring points on the surface, (x_0,y_0,z_0) and (x,y,z). Let

$$dx = x - x_0, \qquad dy = y - y_0, \qquad dz = z - z_0;$$

then

$$\frac{\partial f}{\partial x}\bigg|_{(x_0,y_0,z_0)} (x - x_0) + \frac{\partial f}{\partial y}\bigg|_{(x_0,y_0,z_0)} (y - y_0) + \frac{\partial f}{\partial z}\bigg|_{(x_0,y_0,z_0)} (z - z_0) = 0$$

is the equation of a plane through (x_0,y_0,z_0), as the partial derivatives are constants. Let this plane be cut by the plane $x = x_0$. The line of intersection is

$$\frac{\partial f}{\partial y}\bigg|_{(x_0,y_0,z_0,)} (y - y_0) = -\frac{\partial f}{\partial z}\bigg|_{(x_0,y_0,z_0,)} (z - z_0)$$

which is tangent to the curve $f(x_0,y,z) = 0$ at (x_0,y_0,z_0), since

$$\frac{dz}{dy}\bigg|_{(y_0,z_0)} = -\frac{\partial f/\partial y|_{(x_0,y_0,z_0)}}{\partial f/\partial z|_{(x_0,y_0,z_0)}} = \frac{z - z_0}{y - y_0}$$

Similarly, the planes $y = y_0$ and $z = z_0$ intersect the plane on lines tangent to $f(x,y_0,z) = 0$ and $f(x,y,z_0) = 0$, respectively, at (x_0,y_0,z_0). The plane that contains by substituting for l, m, n in Eq. (11)

$$\dot{\nu} = -\frac{1}{R}\frac{\partial F}{\partial t} \tag{13}$$

Substitution of Eq. (13) into Eq. (7) using Eq. (12) gives

$$u\frac{\partial F}{\partial x} + v\frac{\partial F}{\partial y} + w\frac{\partial F}{\partial z} + \frac{\partial F}{\partial t} = 0 \tag{14}$$

which is the general boundary condition equation. This may be written

$$\frac{DF}{Dt} = 0 \tag{15}$$

where the operator $\dfrac{D}{Dt}$ is

$$\frac{D}{Dt} = u\frac{\partial}{\partial x} + v\frac{\partial}{\partial y} + w\frac{\partial}{\partial z} + \frac{\partial}{\partial t} \tag{16}$$

This operation is called differentiation with respect to the motion; thus

$$\frac{Du}{Dt} = a_x, \qquad \frac{Dv}{Dt} = a_y, \qquad \frac{Dw}{Dt} = a_z$$

Example: The equation of a sphere of radius r whose center is moving along the x-axis with velocity U is given by

$$F = (x - Ut)^2 + y^2 + z^2 - r^2 = 0$$

The boundary condition is found as follows:

$$\frac{DF}{Dt} = 0 \qquad \text{or} \qquad u\frac{\partial F}{\partial x} + v\frac{\partial F}{\partial y} + w\frac{\partial F}{\partial z} + \frac{\partial F}{\partial t} = 0$$

and

$$\frac{\partial F}{\partial x} = 2(x - Ut), \qquad \frac{\partial F}{\partial y} = 2y, \qquad \frac{\partial F}{\partial z} = 2z, \qquad \frac{\partial F}{\partial t} = -2U(x - Ut)$$

---

these three tangent lines at (x_0, y_0, z_0) is defined as the tangent plane to the surface at (x_0, y_0, z_0).

The direction cosines of the normal to the tangent plane are proportional to the coefficients of x, y, z in the equation of the plane,

$$\frac{\partial f}{\partial x}\bigg|_{(x_0,y_0,z_0)} : \frac{\partial f}{\partial y}\bigg|_{(x_0,y_0,z_0)} : \frac{\partial f}{\partial z}\bigg|_{(x_0,y_0,z_0)}$$

and the direction cosines of the normal to the surface at (x_0, y_0, z_0) are

$$l = \frac{\partial f/\partial x}{R}, \qquad m = \frac{\partial f/\partial y}{R}, \qquad n = \frac{\partial f/\partial z}{R}$$

where

$$R = \sqrt{\left(\frac{\partial f}{\partial x}\right)^2 + \left(\frac{\partial f}{\partial y}\right)^2 + \left(\frac{\partial f}{\partial z}\right)^2}$$

Hence,

$$\frac{DF}{Dt} = u(x - Ut) + vy + wz - U(x - Ut) = 0$$

or

$$(u - U)(x - UT) + vy + wz = 0$$

This is the required boundary condition equation which must be satisfied for every point on the surface of the sphere.

In dealing with the motion of bodies through an infinite fluid, it is assumed in general that the motion starts from rest; *i.e.*, initially both fluid and bodies are stationary. When forces are applied to the bodies to set them in motion, a necessary condition is that the fluid at infinity remain at rest. If this were not so, it would imply that the finite forces acting on the bodies had imparted to the fluid infinite kinetic energy in finite time, which is obviously impossible.

Dynamical boundary conditions may also arise when two different fluids are in contact. The pressure intensity must vary continuously across the boundary. If there were a finite change in pressure across the boundary, then the resultant force produced on an infinitesimal slice of the boundary would produce infinite acceleration of the boundary.

9. Irrotational Flow. Velocity Potential. A further restriction to the general fluid problem, *viz.*, the assumption that the flow is irrotational, provides a means of integrating the Euler equations, if the extraneous (body) forces are derivable from a potential. It is shown in this section that the assumption of irrotational flow is identical with the assumption that a velocity potential exists.

Rotation may be represented as a vector that has three components parallel to the *xyz*-axes. The sense of rotation follows the right-handed rule; for example, the component extending in the positive direction along the *x*-axis indicates rotation about an axis parallel to *x* in the sense that would make a right-handed screw progress in the positive *x*-direction. The length of the vector (or component) is a measure of the magnitude of the rotation (radians per second).

The rotation component of a fluid particle about an axis through itself, say parallel to the *z*-axis, may be defined as the average angular velocity of any two infinitesimal linear elements in the particle that are perpendicular to each other and to the *z*-axis. The rotation components are given the symbol ω with subscripts indicating the axis to which they are parallel. Hence, ω_x, ω_y, ω_z are the components of rotation about axes parallel to the *xyz*-axes of the coordinate system, respectively; they may be added vectorially to find the total rotation of the particle. The magnitude of the rotation is given by

$$\omega = \sqrt{\omega_x{}^2 + \omega_y{}^2 + \omega_z{}^2}$$

The rotation vector of a particle is defined as one-half the curl of the velocity vector.

Returning to the definition of rotation, the mathematical expression for the rotation component of a particle about an axis parallel to z is developed. Any two line segments perpendicular to z through the particle will be in a plane parallel to the xy-plane. These two lines may conveniently be taken parallel to the x- and y-axes, as they must be at right angles to each other, although any other two perpendicular lines in the plane through the point would give the same result. Referring to Fig. 7, the particle is at $P(x,y,z)$ and has the velocity components u,v in the xy-plane. The angular velocities of δx and δy are sought. The angular velocity of the δx segment is

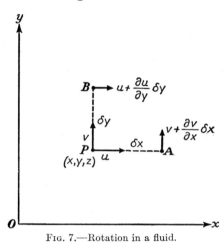

Fig. 7.—Rotation in a fluid.

$$\frac{v + \frac{\partial v}{\partial x}\,\delta x - v}{\delta x} = \frac{\partial v}{\partial x} \quad \text{radians per second}$$

The angular velocity of the δy segment is

$$\frac{u + \frac{\partial u}{\partial y}\,\delta y - u}{\partial y} = -\frac{\partial u}{\partial y} \quad \text{radians per second}$$

using counterclockwise as positive.

Hence, by definition, the rotation of the particle situated at (x,y,z) about an axis parallel to the z-axis is

$$\omega_z = \frac{1}{2}\left(\frac{\partial v}{\partial x} - \frac{\partial u}{\partial y}\right)$$

The rotation components about axes parallel to x and y are found in an analogous manner to be

$$\omega_x = \frac{1}{2}\left(\frac{\partial w}{\partial y} - \frac{\partial v}{\partial z}\right), \qquad \omega_y = \frac{1}{2}\left(\frac{\partial u}{\partial z} - \frac{\partial w}{\partial x}\right)$$

Now that rotation has been defined, irrotational flow may be defined as the absence of rotation at every point in the fluid with the exception of singular points.   That is,

$$\omega_x = \omega_y = \omega_z = 0$$

or

$$\frac{\partial w}{\partial y} = \frac{\partial v}{\partial z}, \qquad \frac{\partial u}{\partial z} = \frac{\partial w}{\partial x}, \qquad \frac{\partial v}{\partial x} = \frac{\partial u}{\partial y} \qquad (17)$$

must be satisfied throughout the fluid.

A visual concept of irrotational flow may be obtained by considering as a free body a small element of fluid in the shape of a sphere.   As the fluid is frictionless, no tangential stresses or forces may be applied to its surface.   The pressure forces act normal to the surface and, hence, through its center.   Extraneous forces act through its mass center, which is also its geometric center for constant density.   Hence, it is evident that no torque may be applied about any diameter of the sphere.   Therefore, the angular acceleration of the sphere about any axis through the sphere must always be zero.   If the sphere is initially at rest, it cannot be set in rotation by any means whatsoever; and if initially in rotation, there is no means of changing its rotation.   As this applies to every point in the fluid, except singular points, one may visualize the fluid elements as being pushed around by boundary movements but not being rotated if initially at rest.   Rotation or lack of rotation of the fluid particles is a property of the fluid itself and not its position in space.   For example, a certain space may be occupied by irrotational fluid at one time, and at some later time other fluid with rotation may have taken its place.

In an actual fluid, such as air or water, having small viscosity, irrotational flow may take place for all practical purposes for a short time after motion starts from rest.

The concept of a velocity potential ϕ in fluid flow has several advantages.   It is defined as a scalar function of space and time such that its negative derivative with respect to any direction is the fluid velocity in that direction.   It is analogous to the force potential whose derivative with respect to a direction is the force in that direction.   Mathematically the velocity potential ϕ is defined by the equations

$$u = -\frac{\partial \phi}{\partial x} \qquad (18a)$$

$$v = -\frac{\partial \phi}{\partial y} \qquad (18b)$$

$$w = -\frac{\partial \phi}{\partial z} \qquad (18c)$$

where u, v, w are the component velocities in the xyz-directions.   By means of Eqs. (18) partial derivatives of ϕ may replace the velocity components u, v, and w.

As

$$\frac{\partial^2 \phi}{\partial x \, \partial y} = \frac{\partial^2 \phi}{\partial y \, \partial x}$$

differentiation of Eq. (18a) with respect to y and of Eq. (18b) with respect to x shows that

$$\frac{\partial u}{\partial y} = \frac{\partial v}{\partial x}$$

In a similar manner

$$\frac{\partial u}{\partial z} = \frac{\partial w}{\partial x} \quad \text{and} \quad \frac{\partial v}{\partial z} = \frac{\partial w}{\partial y}$$

which are the conditions for irrotational flow [Eqs. (17)] derived in this section.

Thus, it is seen that the existence of a velocity potential implies irrotational flow. To show that the conditions for irrotational flow imply the existence of a velocity potential, the equation

$$\frac{\partial u}{\partial y} = \frac{\partial v}{\partial x}$$

is the necessary and sufficient condition that

$$u \, dx + v \, dy$$

is a perfect differential, say $-d\phi_1$, and

$$d\phi_1 = \frac{\partial \phi_1}{\partial x} dx + \frac{\partial \phi_1}{\partial y} dy$$

ϕ_1 is a function of x,y,z, where z is considered constant.

Next, consider the function

$$\phi(x,y,z) = \phi_1(x,y,z) + f(z)$$

with $f(z)$ a function of z only. Since

$$\frac{\partial \phi}{\partial x} = \frac{\partial \phi_1}{\partial x}, \qquad \frac{\partial \phi}{\partial y} = \frac{\partial \phi_1}{\partial y}$$

it is desired to find $f(z)$ such that ϕ is the velocity potential, that is,

$$\frac{\partial \phi}{\partial z} = \frac{\partial \phi_1}{\partial z} + f'(z) = -w$$

where $f'(z)$ is $\dfrac{\partial f(z)}{\partial z}$. Then

$$f'(z) = -w - \frac{\partial \phi_1}{\partial z}$$

Partial derivatives of $f'(z)$ with respect to x and y must be zero, since it is a function of z only, thus

$$\frac{\partial f'(z)}{\partial x} = -\frac{\partial w}{\partial x} - \frac{\partial}{\partial x}\frac{\partial \phi_1}{\partial z} = -\frac{\partial w}{\partial x} - \frac{\partial}{\partial z}\frac{\partial \phi_1}{\partial x} = -\frac{\partial w}{\partial x} + \frac{\partial u}{\partial z} = 0$$

$$\frac{\partial f'(z)}{\partial y} = -\frac{\partial w}{\partial y} - \frac{\partial}{\partial y}\frac{\partial \phi_1}{\partial z} = -\frac{\partial w}{\partial y} - \frac{\partial}{\partial z}\frac{\partial \phi_1}{\partial y} = -\frac{\partial w}{\partial y} + \frac{\partial v}{\partial z} = 0$$

which are Eqs. (17). Hence $f'(z)$ is determined, from which $f(z)$ and ϕ may be determined.

Hence, the conditions for irrotational flow give rise to the defining equations for velocity potential. In these relations the partial derivatives are assumed to be continuous. Therefore, the assumption of irrotational flow and the assumption that a velocity potential exists are one and the same thing.

Using the relationships obtained by assuming the existence of a velocity potential, the Euler equations may now be integrated if the extraneous forces are derivable from a potential. Gravity is the usual extraneous force encountered in fluid flow and is derivable from a potential.

10. Integration of Euler's Equations. Bernoulli Equation. If the extraneous force components are given by

$$X = -\frac{\partial \Omega}{\partial x}, \qquad Y = -\frac{\partial \Omega}{\partial y}, \qquad Z = -\frac{\partial \Omega}{\partial z}$$

where Ω is the extraneous force potential, the first of Euler's equations, Eqs. (3), may be written

$$u\frac{\partial u}{\partial x} + v\frac{\partial u}{\partial y} + w\frac{\partial u}{\partial z} + \frac{\partial u}{\partial t} + \frac{\partial \Omega}{\partial x} + \frac{1}{\rho}\frac{\partial p}{\partial x} = 0$$

Substituting in the irrotational flow conditions

$$\frac{\partial u}{\partial y} = \frac{\partial v}{\partial x}, \qquad \frac{\partial u}{\partial z} = \frac{\partial w}{\partial x}$$

and the velocity potential

$$u = -\frac{\partial \phi}{\partial x}$$

the equation takes the form

$$u\frac{\partial u}{\partial x} + v\frac{\partial v}{\partial x} + w\frac{\partial w}{\partial x} - \frac{\partial^2 \phi}{\partial t\,\partial x} + \frac{\partial \Omega}{\partial x} + \frac{1}{\rho}\frac{\partial p}{\partial x} = 0$$

As

$$u\frac{\partial u}{\partial x} = \frac{\partial}{\partial x}\left(\frac{u^2}{2}\right), \qquad v\frac{\partial v}{\partial x} = \frac{\partial}{\partial x}\left(\frac{v^2}{2}\right), \qquad w\frac{\partial w}{\partial x} = \frac{\partial}{\partial x}\left(\frac{w^2}{2}\right)$$

the equation may be expressed as

$$\frac{\partial}{\partial x}\left(\frac{u^2}{2} + \frac{v^2}{2} + \frac{w^2}{2} - \frac{\partial \phi}{\partial t} + \Omega + \frac{p}{\rho}\right) = 0$$

where ρ is now considered a constant. Integrating with respect to x

$$\frac{u^2 + v^2 + w^2}{2} - \frac{\partial \phi}{\partial t} + \Omega + \frac{p}{\rho} = F_1(y,z,t)$$

where F_1 is an arbitrary function resulting from the integration. Defining q as the magnitude of the velocity, or the speed, at any point, then, as

$$q^2 = u^2 + v^2 + w^2$$

the integrated equation is

$$\tfrac{1}{2}q^2 - \frac{\partial \phi}{\partial t} + \Omega + \frac{p}{\rho} = F_1(y,z,t)$$

Integrating the second and third Euler equations in an analogous manner,

$$\tfrac{1}{2}q^2 - \frac{\partial \phi}{\partial t} + \Omega + \frac{p}{\rho} = F_2(x,z,t)$$

and

$$\tfrac{1}{2}q^2 - \frac{\partial \phi}{\partial t} + \Omega + \frac{p}{\rho} = F_3(x,y,t)$$

As the left-hand sides of the equations are the same

$$F_1(y,z,t) = F_2(x,z,t) = F_3(x,y,t)$$

where the functions are all arbitrary. Examining the first two members, since x and y are independent variables, they must disappear from the equation or the equation could not be true. Similarly, considering the first and third members, z must disappear from the functions. F_1, F_2, and F_3 must then reduce to an arbitrary function of the time alone, or to a constant. The final integrated form of the three Euler differential equations becomes one equation:

$$\tfrac{1}{2}q^2 - \frac{\partial \phi}{\partial t} + \Omega + \frac{p}{\rho} = F(t) \tag{19}$$

containing an arbitrary function of time $F(t)$. This is the Bernoulli equation for unsteady flow, *i.e.*, for flow that changes with time. In steady flow where there is no change in conditions with respect to time, the equation reduces to

$$\tfrac{1}{2}q^2 + \Omega + \frac{p}{\rho} = C \tag{20}$$

where C is an arbitrary constant to be determined by known conditions of velocity, pressure, and extraneous force potential at some point in the flow.

Restricting Euler's equation to steady flow along a streamline with q the total velocity, the equation becomes

$$q\frac{\partial q}{\partial s} + \frac{\partial \Omega}{\partial s} + \frac{1}{\rho}\frac{\partial p}{\partial s} = 0$$

where s is the distance measured along the streamline. Integrating with respect to s,

$$\frac{q^2}{2} + \Omega + \frac{p}{\rho} = C$$

where C is a constant for a particular streamline. This derivation does not require irrotational flow but requires that all points to which the equation applies be on the same streamline.

The units of each term in Eqs. (19) and (20) are energy per unit mass (foot-pound per slug or square feet per second per second). Each term on the left-hand side of Eq. (20) may be considered a form of energy (kinetic, potential, and pressure energy, respectively), and their sum is a constant throughout the whole region of steady irrotational flow.

Orienting the z-axis positive upward and making the additional assumption that gravity is the only extraneous force acting, then

$$\Omega = +gz$$

It is convenient to consider the pressure intensity as made up of two parts, that due to static fluid conditions p_s, the pressure that would exist if there were no motion, and that due to dynamic conditions p_d, *i.e.*, to changes in velocity. Equation (19) may be written

$$\tfrac{1}{2}q^2 - \frac{\partial \phi}{\partial t} + gz + \frac{p_s}{\rho} + \frac{p_d}{\rho} = F(t)$$

The two terms

$$\frac{p_s}{\rho} + gz$$

equal a constant, however, as an increase in potential energy due to increase in z is exactly offset by a decrease in static pressure energy. These two terms may then be included in $F(t)$, reducing the formula to

$$\tfrac{1}{2}q^2 - \frac{\partial \phi}{\partial t} + \frac{p}{\rho} = F(t) \tag{21}$$

where p now becomes the dynamic pressure. Its subscript is dropped; hence it should be remembered that the total pressure at a point is found by adding to the p of Eq. (21) (and subsequent equations) the static pressure, which is the pressure that would exist if there were no flow.

For steady flow, Eq. (21) reduces to

$$\tfrac{1}{2}q^2 + \frac{p}{\rho} = C$$

The constant C in this equation is usually determined by known conditions for velocity and dynamic pressure at some point in the flow. Thus, if the velocity of the undisturbed flow is q_0 and the dynamic pressure is zero, C has the value $\tfrac{1}{2}q_0^2$. Substituting in this value for C, the pressure intensity at any other point is given by

$$p = \frac{\rho}{2}(q_0^2 - q^2) \tag{22}$$

Euler's equations have been integrated without using the continuity equation or any particular boundary conditions. As the continuity equation and the boundary conditions are both kinematical (contain no density term), the magnitude and direction of velocity at all points are independent of the particular fluid, and the problem resolves itself into one of geometry—to find the particular velocity potential that satisfies continuity and the prescribed boundary conditions. Once the velocity distribution (or field) is known, the pressure may be determined from the Bernoulli equation.

11. The Laplace Equation. Equipotential Surfaces. When the conditions for existence of a velocity potential (equivalent to assumption of irrotational flow) [Eqs. (18)] are substituted into the continuity equation, Eq. (5), the Laplace equation results:

$$\frac{\partial^2\phi}{\partial x^2} + \frac{\partial^2\phi}{\partial y^2} + \frac{\partial^2\phi}{\partial z^2} = 0 \tag{23}$$

which must be satisfied at every point throughout the fluid, excepting singular points. Equation (23) is usually written in symbolic form

$$\nabla^2\phi = 0 \tag{24}$$

where ∇^2 (pronounced "del squared") is the operator

$$\nabla^2 = \frac{\partial^2}{\partial x^2} + \frac{\partial^2}{\partial y^2} + \frac{\partial^2}{\partial z^2} \tag{25}$$

when expressed in cartesian coordinates. The Laplace equation is encountered in many other branches of physics, such as electricity, heat flow, and elasticity. It has the units T^{-1} (per second). As the Laplace equation contains the continuity equation and the irrotational flow condition, any function ϕ that satisfies it is a possible fluid flow case. One fruitful method of approach is to examine solutions of the Laplace equation in order to determine which particular boundary conditions they fulfill.

When the boundary conditions are specified in advance, it may be very difficult to find the proper function ϕ to satisfy them as well as the Laplace equation. Several methods of attack are available other than

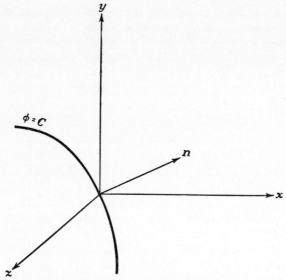

Fig. 8.—Normal to an equipotential surface.

the direct method of finding ϕ from the boundary condition equations. These include the electric analogy method, the Hele Shaw method, and graphical trial-and-error methods.

Equipotential surfaces are defined as those surfaces within the fluid over which the value of ϕ remains constant. As the derivative of ϕ with respect to an element of length in the equipotential surface is zero, *i.e.*, $\dfrac{d\phi}{ds} = 0$, as ϕ is constant along the surface and δs is an element of length in the equipotential surface, there can be no component of the velocity tangential to the surface. In other words, the velocity is everywhere normal to the equipotential surfaces.

The velocity normal to an equipotential surface (Fig. 8) is the total velocity q whose magnitude is given by

$$q = l'u + m'v + n'w = -\left(l'\frac{\partial \phi}{\partial x} + m'\frac{\partial \phi}{\partial y} + n'\frac{\partial \phi}{\partial z} \right) = -\frac{\partial \phi}{\partial n} \quad (26)$$

where δn is an element of the normal to the equipotential surface drawn positive in the direction of decreasing ϕ. l', m', n' are the direction cosines of the normal n. Equation (26) is the negative of the "gradient" of ϕ, written

$$q = - \text{grad } \phi$$

Other surfaces or lines in two-dimensional flow have the velocity vector everywhere tangent to them. They are known as *stream surfaces* or *streamlines*. It follows that stream surfaces and equipotential surfaces intersect orthogonally (at right angles).

12. Summary. The following assumptions have been made in the derivation of Euler's equations, the continuity equation, and the general Bernoulli equation:

1. The fluid is frictionless (viscosity zero).
2. The density is constant.
3. The flow is irrotational; *i.e.*, a velocity potential exists.

The solution of a flow problem was shown to consist of a solution of the Laplace equation that also satisfies the prescribed boundary conditions. The Laplace equation in cartesian coordinates is

$$\frac{\partial^2 \phi}{\partial x^2} + \frac{\partial^2 \phi}{\partial y^2} + \frac{\partial^2 \phi}{\partial z^2} = 0$$

hence, its solution would be an expression for ϕ, the velocity potential, as a function of the coordinates and the time. With the velocity potential known that satisfies the boundary conditions, the velocity field may be obtained from the defining equations for velocity potential; *viz.*,

$$u = - \frac{\partial \phi}{\partial x}, \qquad v = - \frac{\partial \phi}{\partial y}, \qquad w = - \frac{\partial \phi}{\partial z}$$

where u, v, w are the component velocities parallel to the xyz-directions at any point (x,y,z). The pressure distribution throughout the fluid may then be obtained from the appropriate form of the Bernoulli equation.

Practically, the solution of problems in most cases is indirect in that it is much easier to investigate various functions which satisfy the Laplace equation in order to determine which boundary conditions are satisfied by them.

In steady flow a streamline or stream surface may be replaced by a solid boundary without affecting the flow, as the normal component of velocity at a stream surface is always zero when the surface is stationary. Therefore, if any streamline or stream surface can be found that has the exact size, shape, and position of fixed boundaries in a particular problem, then from the potential function the velocity and pressure fields may be obtained.   The assumption of frictionless flow precludes the development of a boundary layer.   In actual problems where a boundary layer may develop appreciable thickness and especially where separation of the flow lines from the boundary may occur, ideal fluid solutions must be used with extreme caution.

Exercises

1. Find the total differential of the following:

(a) $u = \dfrac{x^2 y^2}{zt}$ $\qquad\qquad$ (b) $u = \sin 2x + \cos 2y$

(c) $u = \tan^{-1}(x + y)$ $\qquad\qquad$ (d) $u = \ln(x^2 + xy)$

where ln is the natural logarithm.

2. Given: $x = \sin t,\ y = \ln t,\ z = t + 1$. Find $\dfrac{du}{dt}$ for the functions in Exercise 1.

3. The velocity field is given by

$$u = yz + t, \qquad v = xz - t, \qquad w = xy$$

Find the acceleration components of a fluid particle at $(1,1,1)$ in terms of t.

4. Reduce Euler's equations to the special case where the x-axis (s-axis) is oriented in the direction of the velocity vector at the point (x,y,z) or (s,n,m) and the y-axis (n-axis) is in the direction of the center of curvature of the streamline at the point.   The z-axis (m-axis) then forms a right-handed cartesian system with the other two axes.   Let gravity be the only extraneous force acting with h measured vertically upward. $u_s,\ u_n,\ u_m$ are the velocity components parallel to the snm-axes. $\gamma = \rho g;\ r = $ radius of curvature.

$Ans.$ $\quad -\dfrac{1}{\rho}\dfrac{\partial}{\partial s}(p + \gamma h) = u_s\dfrac{\partial u_s}{\partial s} + \dfrac{\partial u_s}{\partial t};\ -\dfrac{1}{\rho}\dfrac{\partial}{\partial n}(p + \gamma h) = \dfrac{u_s{}^2}{r} + \dfrac{\partial u_n}{\partial t};$

$$-\frac{1}{\rho}\frac{\partial}{\partial m}(p + \gamma h) = \frac{\partial u_m}{\partial t}.$$

5. Which of the following velocity fields are possible fluid flow cases?   Which satisfy the conditions for irrotational flow?

(a) $u = x + y + z,\ v = x - y + z,\ w = x + y + 3$
(b) $u = xy,\ v = yz,\ w = yz + z^2$
(c) $u = x(y + z),\ v = y(x + z),\ w = -(x + y)z - z^2$

(d) $u = xyzt,\ v = -xyzt^2,\ w = \dfrac{z^2}{2}(xt^2 - yt)$

6. With the two velocity components given, find the other component that satisfies continuity.   Are the answers unique?

(a) $u = x^2 + y^2 + z^2$
$\quad v = -xy - yz - xz$
$\quad w =$
(b) $u = \ln(y^2 + z^2)$
$\quad v = \sin(x^2 + z^2)$
$\quad w =$

(c) $u =$

$$v = \frac{+y}{(\sqrt{x^2 + y^2 + z^2})^3}$$

$$w = \frac{+z}{(\sqrt{x^2 + y^2 + z^2})^3}$$

7. Do any of the velocity fields in Exercise 6 satisfy the conditions for potential flow?

8. Write the boundary condition equation for a sphere whose center moves along the positive y-axis with a velocity of 10 ft per sec at the origin and a constant acceleration of 1.0 ft per sec². The radius of the sphere decreases inversely as the distance of its center from $y = -1$ ft and is 1.0 ft when the center is at the origin ($t = 0$).

Ans. $ux + v\left[y - \left(\dfrac{t^2}{2} + 10t\right)\right] + wz$

$$+ (t + 10)\left[\frac{1}{\left(\dfrac{t^2}{2} + 10t + 1\right)^3} - y + \frac{t^2}{2} + 10t\right] = 0.$$

9. Examine the following functions to determine if they could represent the velocity potential for ideal fluid flow.

(a) $f = x + y + z$ (b) $f = x + xy + xyz$
(c) $f = x^2 + y^2 + z^2$ (d) $f = zx^2 - y^2 - z^2$
(e) $f = \sin(x + y + z)$ (f) $f = \ln x$

10. Show that if ϕ_1 and ϕ_2 are both solutions of $\nabla^2 \phi = 0$, then $\phi_1 + \phi_2$, $C\phi_1$, and $C + \phi_1$ are also solutions, where C is a constant.

11. In Sec. 9, the rotation component about an axis parallel to the z-axis was defined as the average angular velocity of two elements, mutually perpendicular to each other and to the z-axis. To prove that the same answer is obtained for any two such elements, show that

$$\frac{1}{2}\left(\frac{\partial v}{\partial x} - \frac{\partial u}{\partial y}\right) = \frac{1}{2}\left(\frac{\partial v'}{\partial x'} - \frac{\partial u'}{\partial y'}\right)$$

where x', y' are a rotated set of axes, the z-axis remaining the same; u', v' are velocity components parallel to x', y', respectively.

CHAPTER III

THEOREMS AND BASIC FLOW DEFINITIONS

13. Equation of Energy. The principle of work and energy is valid for a fluid, provided that it is nonviscous, that the extraneous (or body) forces are derivable from a potential Ω such that

$$X = -\frac{\partial \Omega}{\partial x}, \qquad Y = -\frac{\partial \Omega}{\partial y}, \qquad Z = -\frac{\partial \Omega}{\partial z} \tag{1}$$

and, furthermore, that Ω is independent of time, i.e., $\frac{\partial \Omega}{\partial t} = 0$. In this section it is shown (1) that the time rate of doing work on the boundaries of an ideal fluid is equal to the time rate of change of the sum of the kinetic and potential energies and (2) that the time rate of doing work on the boundaries of a compressible, nonviscous fluid is equal to the time rate of change of the sum of the kinetic, potential, and intrinsic energies.

Euler's equations [Eqs. (3), Sec. 6] are written in the form force per unit mass = acceleration, for the xyz-directions. Multiplying each equation through by $\rho\, \delta x\, \delta y\, \delta z$ makes the equations applicable to an element of mass: resultant force = δ (mass) $\times$ acceleration. Now, by multiplying the x-component by u, the left-hand side is the power applied to the mass particle by forces acting in the x-direction (time rate of doing work), thus

$$uX\rho\, \delta x\, \delta y\, \delta z - u\frac{\partial p}{\partial x}\, \delta x\, \delta y\, \delta z = u\rho\, \delta x\, \delta y\, \delta z\, \frac{Du}{Dt}$$

where the operator $\frac{D}{Dt}$ has the meaning of Eq. (16), Sec. 8. Similarly, by multiplying the y-component by v and the z-component by w and then adding the three equations, the total power added to the particle is obtained:

$$(uX + vY + wZ)\rho\, \delta x\, \delta y\, \delta z - \left(u\frac{\partial p}{\partial x} + v\frac{\partial p}{\partial y} + w\frac{\partial p}{\partial z} \right) \delta x\, \delta y\, \delta z$$

$$= \left(u\frac{Du}{Dt} + v\frac{Dv}{Dt} + w\frac{Dw}{Dt} \right) \rho\, \delta x\, \delta y\, \delta z$$

Substituting in Eqs. (1), noting that $u \dfrac{Du}{Dt} = \dfrac{D}{Dt}\left(\dfrac{u^2}{2}\right)$, etc., the equation takes the form

$$-\frac{D\Omega}{Dt}\rho\,\delta x\,\delta y\,\delta z - \left(u\frac{\partial p}{\partial x} + v\frac{\partial p}{\partial y} + w\frac{\partial p}{\partial z}\right)\delta x\,\delta y\,\delta z$$

$$= \frac{D}{Dt}\left[\frac{\rho}{2}\left(u^2 + v^2 + w^2\right)\delta x\,\delta y\,\delta z\right]$$

Writing $q^2 = u^2 + v^2 + w^2$ and integrating throughout the volume of fluid, this becomes

$$\frac{D}{Dt}\frac{1}{2}\int\int\int \rho q^2\,dx\,dy\,dz + \frac{D}{Dt}\int\int\int \Omega\rho\,dx\,dy\,dz$$

$$= -\int\int\int\left(u\frac{\partial p}{\partial x} + v\frac{\partial p}{\partial y} + w\frac{\partial p}{\partial z}\right)dx\,dy\,dz \quad (2)$$

The first integral

$$T = \tfrac{1}{2}\iiint\rho q^2\,dx\,dy\,dz \tag{3}$$

is the total kinetic energy T in the fluid region. As Ω is the potential energy per unit mass at the point (x,y,z), the second integral is the total potential energy V of the fluid region:

$$V = \iiint\Omega\rho\,dx\,dy\,dz \tag{4}$$

Equation (2) may now be written

$$\frac{D}{Dt}(T + V) = -\int\int\int\left(u\frac{\partial p}{\partial x} + v\frac{\partial p}{\partial y} + w\frac{\partial p}{\partial z}\right)dx\,dy\,dz \tag{5}$$

The integral on the right-hand side of the equation may be integrated by parts. Integrating the x-component of the first term,

$$\int\int\int u\frac{\partial p}{\partial x}\,dx\,dy\,dz = \int\int pu\Big]_{x_1}^{x_2}\,dy\,dz - \int\int\int p\frac{\partial u}{\partial x}\,dx\,dy\,dz$$

where the integration of the first term on the right-hand side of the equation with respect to x is carried out from x_1 to x_2 as in Fig. 9. An element of boundary surface δS is related to its projection on the yz-plane, $\delta y\,\delta z$, by

$$\delta y\,\delta z = \pm l\,\delta S$$

where l, m, n are the direction cosines of the normal to the surface, drawn into the fluid. As δy, δz, and δS are always considered positive, the minus sign is used when l is negative. Substituting in the value of $dy\,dz$ at x_2 and x_1, it is evident that

$$\int \int pu \left. \right]_{x_1}^{x_2} dy\, dz = - \int pul\, dS$$

where the integration is carried out over the whole boundary surface of the fluid. Transforming the remaining terms in an analogous manner,

$$\frac{D}{Dt}(T + V) = \int p(lu + mv + nw)dS$$

$$+ \int \int \int p \left(\frac{\partial u}{\partial x} + \frac{\partial v}{\partial y} + \frac{\partial w}{\partial z} \right) dx\, dy\, dz \quad (6)$$

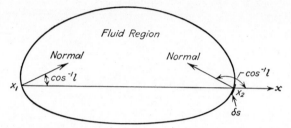

FIG. 9.—Fluid region showing the direction cosine of a normal to the boundary.

For an incompressible fluid the integrand of the last term is zero, by Eq. (5), Sec. 7. Hence,

$$\frac{D}{Dt}(T + V) = \int p(lu + mv + nw)\, dS \quad (7)$$

Since $lu + mv + nw$ is the velocity of the boundary normal to itself [Eq. (7), Sec. 8] and $p\, dS$ is the force element exerted on the fluid by the boundary, the surface integral is the rate at which work is being done on the fluid by forces at the boundary. For fixed boundaries, Eq. (7) reduces to

$$T + V = \text{constant}$$

For a compressible, nonviscous fluid the right-hand term of Eq. (6) may be interpreted as the time rate of change of intrinsic energy of the fluid in the region. Intrinsic energy may be defined as the work that a fluid is capable of doing by expanding against external pressure. The terms

$$\frac{\partial u}{\partial x} + \frac{\partial v}{\partial y} + \frac{\partial w}{\partial z}$$

represent the expansion rate per unit volume of fluid. Multiplying by the pressure intensity p, the work done per unit time per unit volume is obtained. Multiplying this by $\delta x\, \delta y\, \delta z$ and integrating throughout the region give the time rate of work done by the expanding fluid, or the time rate of decrease in intrinsic energy. Let W represent the intrinsic energy of the fluid region; then

$$-\frac{DW}{Dt} = \int \int \int p\left(\frac{\partial u}{\partial x} + \frac{\partial v}{\partial y} + \frac{\partial w}{\partial z}\right) dx\, dy\, dz$$

Equation (6) now becomes

$$\frac{D}{Dt}(T + V + W) = \int p(lu + mv + nw)\, dS \tag{8}$$

Therefore, the time rate of doing work on the boundaries is equal to the time rate of change of fluid energy (kinetic, potential, and intrinsic).

No restrictions have been placed on the nature of the fluid flow, *i.e.*, whether it is rotational or irrotational. Therefore, equations derived in this section are valid for either flow.

14. Green's Theorem. Kinetic Energy Equation. The use of Green's theorem permits the derivation of an expression for the kinetic energy of a fluid in irrotational motion. This kinetic energy expression, in turn, aids in establishing some important uniqueness theorems which are given in Sec. 15.

Green's theorem states that

$$\int (lU + mV + nW)\, dS = -\int \int \int \left(\frac{\partial U}{\partial x} + \frac{\partial V}{\partial y} + \frac{\partial W}{\partial z}\right) dx\, dy\, dz \tag{9}$$

where U, V, W are any finite, single-valued, differentiable functions of space in a connected region completely bounded by one or more closed surfaces S, of which δS is an element and l, m, n are the direction cosines of the normal to the surface element δS directed into the region. The double integration is carried out over the boundary, and the triple integration throughout the region.

Consider the region as comprised of a large number of small parallelepipeds with edges $\delta x\ \delta y\ \delta z$, respectively parallel to the xyz-axes of a cartesian system of reference. The surface integral

$$\int (lU + mV + nW)\, dS \tag{10}$$

when applied to the interface between two of the volume elements, will have the normals extending in opposite directions; thus if l_1, m_1, n_1 refer to the inwardly drawn normal from the face for one volume element and l_2, m_2, n_2 refer to the inwardly drawn normal from the same face to the other volume element, $l_1 = -l_2$, $m_1 = -m_2$, $n_1 = -n_2$, and the value for the face is zero, as U, V, W are functions of space. In a similar manner the surface integral becomes zero for all the inner surfaces and hence need be evaluated only for the original boundary of the region.

Consider one of the volume elements with center at (x,y,z).   For the yz-face nearer to the origin the surface integral (10) becomes

$$\left(U - \frac{\partial U}{\partial x} \frac{\delta x}{2} \right) \delta y \; \delta z$$

as $l = 1, m = 0, n = 0, \delta S = \delta y \; \delta z$.   For the opposite face (10) becomes

$$-\left(U + \frac{\partial U}{\partial x} \frac{\delta x}{2} \right) \delta y \; \delta z$$

as $l = -1, m = 0, n = 0, \delta S = \delta y \; \delta z$.   Their sum is

$$-\frac{\partial U}{\partial x} \; \delta x \; \delta y \; \delta z$$

Evaluating (10) for the other two pairs of faces and adding,

$$-\left(\frac{\partial U}{\partial x} + \frac{\partial V}{\partial y} + \frac{\partial W}{\partial z} \right) \delta x \; \delta y \; \delta z$$

is the value of (10) for this element.   Summing up all the surface elements of the region results in the triple integral on the right-hand side of Eq. (9); and as the surface integral over all the elemental volumes equals the surface integral over the original boundary, Green's theorem [Eq. (9)] is proved.

To obtain the expression for kinetic energy let

$$U = \phi \frac{\partial \phi}{\partial x}, \qquad V = \phi \frac{\partial \phi}{\partial y}, \qquad W = \phi \frac{\partial \phi}{\partial z}$$

where ϕ is the velocity potential.   Substituting into Eq. (9), first

$$lU + mV + nW = \phi \left(l \frac{\partial \phi}{\partial x} + m \frac{\partial \phi}{\partial y} + n \frac{\partial \phi}{\partial z} \right) = \phi \frac{\partial \phi}{\partial n}$$

where δn is an element of the normal to the boundary and l, m, n are the direction cosines of the normal.   Then

$$\frac{\partial U}{\partial x} + \frac{\partial V}{\partial y} + \frac{\partial W}{\partial z} = \left(\frac{\partial \phi}{\partial x} \right)^2 + \left(\frac{\partial \phi}{\partial y} \right)^2 + \left(\frac{\partial \phi}{\partial z} \right)^2 + \phi \; \nabla^2 \phi$$

using the notation of Sec. 11.   Multiplying by $\rho/2$ and substituting into Green's theorem,

$$-\frac{\rho}{2} \int \phi \frac{\partial \phi}{\partial n} \, dS = \frac{\rho}{2} \int \int \int \left[\left(\frac{\partial \phi}{\partial x} \right)^2 + \left(\frac{\partial \phi}{\partial y} \right)^2 + \left(\frac{\partial \phi}{\partial z} \right)^2 \right] dx \, dy \, dz$$

$$+ \frac{\rho}{2} \int \int \int \phi \; \nabla^2 \phi \, dx \, dy \, dz$$

For any possible irrotational fluid flow case with constant density, $\nabla^2\phi = 0$ and the last integral drops out of the equation. The remaining triple integral is that of Eq. (3), Sec. 13, the total kinetic energy of the fluid region. Hence, the kinetic energy T of the fluid is

$$T = -\frac{\rho}{2} \int \phi \frac{\partial\phi}{\partial n} dS \tag{11}$$

where the integral is evaluated over the boundary.

This formula may be shown to hold for an infinite region,[1] at rest at infinity, when bounded internally by a solid. Let Σ represent a large surface enclosing the fluid region about the solid S. Then for this region

$$T_\Sigma = -\frac{\rho}{2} \int_{(S)} \phi \frac{\partial\phi}{\partial n} dS - \frac{\rho}{2} \int_{(\Sigma)} \phi \frac{\partial\phi}{\partial n} d\Sigma \tag{12}$$

is the kinetic energy of the fluid. As there is no flow across the face S of the solid, the continuity equation may take the form of a statement that the flow out of the boundaries S and Σ must be zero; thus

$$\int_{(S)} \frac{\partial\phi}{\partial n} dS + \int_{(\Sigma)} \frac{\partial\phi}{\partial n} d\Sigma = 0 \tag{13}$$

Multiplying Eq. (13) by an arbitrary constant $C\rho/2$ and adding to Eq. (12),

$$T_\Sigma = -\frac{\rho}{2} \int_{(S)} (\phi - C) \frac{\partial\phi}{\partial n} dS - \frac{\rho}{2} \int_{(\Sigma)} (\phi - C) \frac{\partial\phi}{\partial n} d\Sigma \tag{14}$$

As the liquid is at rest at infinity, ϕ becomes a constant at infinity. Let C be this constant, and enlarge the surface Σ indefinitely in all directions. The second integral then becomes zero, as $(\phi - C)$ is of order $\frac{1}{R}$, $\frac{\partial\phi}{\partial n}$ is of order $\frac{1}{R^2}$ at great distances from the solid S, while the surface element is of order R^2, R being the distance of the surface from the solid. Hence, as R becomes infinite, the integral vanishes.[2] The first integral reduces to Eq. (11), as

$$C \int_{(S)} \frac{\partial\phi}{\partial n} dS = 0$$

states that there is no flow out of the region through S.

[1] Milne-Thomson, "Theoretical Hydrodynamics," pp. 88–89, Macmillan & Co., Ltd., London, 1938.

[2] The order of ϕ and $\frac{\partial\phi}{\partial n}$ are obtained from considerations of source flow. It may be shown that the motion of a fluid due to the passage of a body through it can be obtained by a suitable distribution of sources, sinks, and doublets about the boundary of the solid. See Lamb, "Hydrodynamics," 6th ed., pp. 58–59, Cambridge University Press, London, 1932.

15. Uniqueness Theorems. The following theorems are proved as a consequence of Eq. (11). These theorems are limited to ideal fluid flow cases where the velocity potential is single-valued.

Irrotational motion of a fluid is impossible if the boundaries are fixed. In

$$ T = \tfrac{1}{2}\rho \int \int \int q^2 \, dx \, dy \, dz = -\tfrac{1}{2}\rho \int \phi \frac{\partial \phi}{\partial n} \, dS \qquad (15) $$

$\dfrac{\partial \phi}{\partial n}$ is zero over all the boundaries (Sec. 8). Hence, $T = 0$, and q, the velocity, is everywhere zero.

Irrotational motion of a fluid will cease when the boundaries come to rest. As $\dfrac{\partial \phi}{\partial n}$ must be zero the instant the boundaries cease to move, q must be zero everywhere to satisfy the kinetic energy equation.

Irrotational motion of a fluid that satisfies the Laplace equation and prescribed boundary conditions is uniquely determined by the motion of the boundaries. Let ϕ_1, ϕ_2 be two solutions that satisfy $\nabla^2 \phi = 0$ and the boundary conditions. Then

$$ \frac{\partial \phi_1}{\partial n} = \frac{\partial \phi_2}{\partial n} $$

at each point on the boundary. As $\phi = \phi_1 - \phi_2$ is also a solution[1] of the Laplace equation,

$$ \frac{\partial \phi}{\partial n} = \frac{\partial \phi_1}{\partial n} - \frac{\partial \phi_2}{\partial n} = \frac{\partial(\phi_1 - \phi_2)}{\partial n} = 0 $$

shows that from Eq. (15) $q = 0$ at every point. Then $\phi = $ constant, and ϕ_1 and ϕ_2 may differ by only a constant. The velocity distribution given by ϕ_1 and ϕ_2 are identical, as ϕ is always subject to the addition of an arbitrary constant without affecting the Laplace equation or boundary conditions.

Irrotational motion of a fluid at rest at infinity is impossible if the interior boundaries are at rest. The proof is essentially the same as for the first theorem above.

Irrotational motion of a fluid at rest at infinity is uniquely determined by the motion of the interior solid boundaries. Let ϕ_1, ϕ_2 be two solutions satisfying the Laplace equation and the boundary conditions. Then

$$ \frac{\partial \phi_1}{\partial n} = \frac{\partial \phi_2}{\partial n} $$

over the solid surface, and $q_1 = q_2 = 0$ at infinity. As $\phi = \phi_1 - \phi_2$ is

---

[1] As $\nabla^2 \phi = 0$ is a linear, homogeneous differential equation, the sum of any two solutions is also a solution. See Exercise 10 at end of Chap. II.

also a solution of $\nabla^2\phi = 0$, $\dfrac{\partial\phi}{\partial n} = 0$ over the surface of the solid. Hence, by Eq. (15), $q = 0$ everywhere, $\phi =$ constant, and ϕ_1 and ϕ_2 produce the same fluid motion.

16. The Stream Function in Two-dimensional Flow. The stream function requires a different definition for two-dimensional flow from that

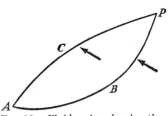

for three-dimensional flow. A streamline, however, is defined in the same manner for either two- or three-dimensional flow, *viz.*, a continuous line through the fluid such that it has the direction of the velocity at every point throughout its length. The differential equation for a streamline is

Fig. 10.—Fluid region showing the positive flow direction used in the definition of a stream function.

$$\frac{dx}{u} = \frac{dy}{v} = \frac{dz}{w}$$

in three-dimensional flow. The last part of the equality, $\dfrac{dz}{w}$, drops out in two-dimensional flow.

In two-dimensional flow all lines of motion are parallel to a fixed plane, say the xy-plane, and the flow patterns (networks of equipotential lines and streamlines) in all planes parallel to this plane are identical.

Let A, P represent two points in one of the planes (Fig. 10), and consider that the flow has unit thickness; *i.e.*, the flow is between two planes, say $z = 0$ and $z = 1$. The rate of flow of fluid across any two lines ACP, ABP must be the same if the density is constant and no fluid is created or destroyed within the region, as a consequence of continuity. Now consider A a fixed point in the xy-plane and P a movable point. The flow rate across any line connecting the two points is a function of the position of P and of the time. Let this function be ψ, and take as sign convention that it denotes the flow rate from right to left across any line connecting A and P, when the observer is at A looking along the line toward P. Thus

$$\psi = \psi(x,y,t)$$

is defined as the stream function.

Let ψ_1, ψ_2 represent the values of the stream function at points P_1, P_2, respectively, in Fig. 11. Then $\psi_2 - \psi_1$ is the flow across P_1P_2 from right to left. If some other point besides A were taken as fixed point, say O, then the values of ψ_1 and ψ_2 would be increased by the same amount, *viz.*, the flow across $\overline{OA}$. ψ is then indeterminate to the extent of an arbitrary constant.

Useful expressions for velocity components may be worked out from the definition of ψ. Displace P in Fig. 12 to P' an infinitesimal distance δy in the y-direction. The flow rate across δy from right to left is $\delta \psi = -u\,\delta y$, or

$$u = -\frac{\partial \psi}{\partial y} \qquad (16)$$

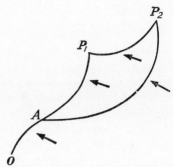

Similarly, if P is displaced an infinitesimal distance δx in the positive x-direction, $\delta \psi = v\,\delta x$ or

$$v = \frac{\partial \psi}{\partial x} \qquad (17)$$

In words, the partial derivative of the stream function with respect to any direction gives the velocity component +90 deg (counterclockwise) to that direction.

Fig. 11.—Flow between two points in a fluid region.

Equations (16) and (17) are true whether the flow has rotation or not. For irrotational flow, however, the condition

$$\frac{\partial u}{\partial y} = \frac{\partial v}{\partial x}$$

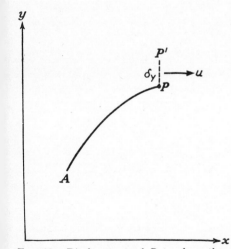

from Eqs. (17), Sec. 9, applies. Substituting Eqs. (16) and (17) into this expression gives

$$\frac{\partial^2 \psi}{\partial x^2} + \frac{\partial^2 \psi}{\partial y^2} = 0 \qquad (18)$$

showing that ψ may be construed as a velocity potential for some other flow. The stream function has the dimensions $L^2 T^{-1}$ (or cubic feet per second per foot of width), the same as the velocity potential.

When the two points P_1, P_2 of Fig. 11 lie on the same streamline, the rate of flow across AP_1 and AP_2 is the same, as there can be no flow

Fig. 12.—Displacement of P to show the relation between ψ and u.

across a streamline. Then $\psi_1 - \psi_2 = 0$. Therefore, the stream function has a constant value along a streamline. The flow rate between any two streamlines (unit width in z-direction) is given by the difference

of the values of the stream function.   Relations between stream function
and velocity potential are found by equating the expressions for velocity
components:

$$\frac{\partial \phi}{\partial x} = \frac{\partial \psi}{\partial y}, \qquad \frac{\partial \phi}{\partial y} = - \frac{\partial \psi}{\partial x}$$

17. Stokes' Stream Function for Three-dimensional Flow.   The
Stokes' stream function is defined only for those three-dimensional flow
cases which have axial symmetry, *i.e.*, where the flow is in a series of planes
passing through a given line and where the flow pattern is identical
in each of the planes.   The intersection of these planes is the axis of
symmetry.

In any one of these planes through the axis of symmetry select two
points A, P, such that A is fixed and P is variable.   Draw a line connecting AP.   The flow through the surface generated by rotating AP about
the axis of symmetry is a function of the position of P.   Let this function
be $2\pi\psi$, and let the axis of symmetry be the x-axis of a cartesian system of
reference.   Then ψ is a function of x and $\hat{\omega}$, where

$$\hat{\omega} = \sqrt{y^2 + z^2}$$

is the distance from P to the x-axis.   The surfaces $\psi = $ constant are
stream surfaces.

To find the relation between ψ and the velocity components u, v'
parallel to the x-axis and the $\hat{\omega}$-axis (perpendicular to x-axis), respectively,
a similar procedure is employed to that of Sec. 16.   Let PP' be an infini-
tesimal step first parallel to $\hat{\omega}$ and then to x, *i.e.*, $PP' = \delta\hat{\omega}$ and then
$PP' = \delta x$.   The resulting relations between stream function and velocity
are given by

$$-2\pi\hat{\omega}\,\delta\hat{\omega}\,u = 2\pi\,\delta\psi \qquad \text{and} \qquad 2\pi\hat{\omega}\,\delta x\,v' = 2\pi\,\delta\psi$$

Solving for u, v',

$$u = -\frac{1}{\hat{\omega}}\frac{\partial \psi}{\partial \hat{\omega}}, \qquad v' = \frac{1}{\hat{\omega}}\frac{\partial \psi}{\partial x} \qquad (19)$$

The same sign convention is used as in the two-dimensional case.

The relations between stream function and potential function are

$$\frac{\partial \phi}{\partial x} = \frac{1}{\hat{\omega}}\frac{\partial \psi}{\partial \hat{\omega}}, \qquad \frac{\partial \phi}{\partial \hat{\omega}} = -\frac{1}{\hat{\omega}}\frac{\partial \psi}{\partial x} \qquad (20)$$

In three-dimensional flow with axial symmetry ψ has the dimensions
$L^3 T^{-1}$, or volume per unit time.

The stream function is used for flow about bodies of revolution that
are frequently expressed most readily in spherical polar coordinates.

Let r be the distance from the origin and θ be the polar angle; the meridian angle is not needed because of axial symmetry. Referring to Fig. 13a and b,

$$2\pi r \sin \theta \; \delta r \; v_\theta = 2\pi \; \delta \psi$$
$$-2\pi r \sin \theta \; r \; \delta \theta \; v_r = 2\pi \; \delta \psi$$

from which

$$v_\theta = \frac{1}{r \sin \theta} \frac{\partial \psi}{\partial r}, \qquad v_r = -\frac{1}{r^2 \sin \theta} \frac{\partial \psi}{\partial \theta} \qquad (21)$$

and

$$\frac{1}{\sin \theta} \frac{\partial \psi}{\partial \theta} = r^2 \frac{\partial \phi}{\partial r}, \qquad \frac{\partial \psi}{\partial r} = -\sin \theta \frac{\partial \phi}{\partial \theta}$$
$$(22)$$

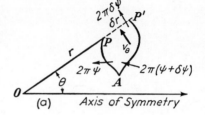

These expressions are useful in dealing with flow about spheres, ellipsoids, and disks and through apertures.

18. Integration of an Exact Equation. The stream function and velocity potential are obtained from perfect differentials. Frequently in dealing with flow cases it is desirable to find the velocity potential if the stream function is known, or vice versa, or to find either the stream function or velocity potential when velocity components are given in terms of the coordinates.

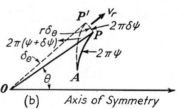

FIG. 13.—Displacement of P to show the relation between velocity components and the stream functions.

Equations (16) to (22) provide means of determining the partial differentials of ϕ or ψ, and by using a straightforward process the values of ϕ and ψ can readily be obtained. This procedure, the integration of an exact differential equation, is given in the following paragraphs.[1]

The equation

$$M \, dx + N \, dy = 0 \qquad (23)$$

is an exact differential du of some function of x and y, say $u(x,y)$, if $\dfrac{\partial M}{\partial y} = \dfrac{\partial N}{\partial x}$. This may be shown as follows:

$$du = \frac{\partial u}{\partial x} \, dx + \frac{\partial u}{\partial y} \, dy$$

is an exact differential (*cf.* Sec. 5); and for Eq. (23) to be exact, the relations

[1] A. Cohen, "Differential Equations," pp. 9–10, D. C. Heath and Company, Boston, 1906.

$$M = \frac{\partial u}{\partial x}, \qquad N = \frac{\partial u}{\partial y}$$

must be true. These are true if $\frac{\partial M}{\partial y} = \frac{\partial N}{\partial x}$, since

$$\frac{\partial^2 u}{\partial y\, \partial x} = \frac{\partial^2 u}{\partial x\, \partial y}$$

Thus the necessity of the condition is proved. To prove its sufficiency and to obtain the solution of the equation, as $M = \frac{\partial u}{\partial x}$,

$$u = \int^x M\, dx + f(y) \qquad (24)$$

where the superscript x on the integral sign indicates integration with respect to x only. $f(y)$ is a function of y only. Since $\frac{\partial u}{\partial y} = N$,

$$\frac{\partial u}{\partial y} = \frac{\partial}{\partial y}\left[\int^x M\, dx + f(y)\right] = N$$

or transposing,

$$\frac{\partial f(y)}{\partial y} = N - \frac{\partial}{\partial y}\int^x M\, dx = \text{a function of } y \text{ alone} \qquad (25)$$

In other words,

$$\frac{\partial}{\partial x}\frac{\partial f(y)}{\partial y} = 0 \qquad \text{or} \qquad \frac{\partial N}{\partial x} - \frac{\partial M}{\partial y} = 0$$

is the condition that $N - \frac{\partial}{\partial y}\int^x M\, dx$ be a function of y alone. This also proves the sufficiency of the condition for an equation to be exact. Proceeding to the solution of the equation by integrating Eq. (25),

$$f(y) = \int \left(N - \frac{\partial}{\partial y}\int^x M\, dx\right) dy$$

Substituting this into Eq. (24) the solution of Eq. (23) is obtained, *viz.*,

$$u = \int^x M\, dx + \int \left(N - \frac{\partial}{\partial y}\int^x M\, dx\right) dy \qquad (26)$$

In words: Integrate $M\, dx$ for y constant. Take the partial derivative with respect to y, and subtract from N, leaving a function of y alone. Integrate this with respect to y, and add to the first integral $\int^x M\, dx$.

As an example of this process, the stream function corresponding to

$$\phi = \frac{\mu}{r^2} \cos \theta$$

is to be determined, where μ is a constant, r and θ are spherical polar coordinates, and the velocity potential has axial symmetry. From Eq. (22), $\frac{\partial \psi}{\partial r}$ and $\frac{\partial \psi}{\partial \theta}$ are obtained; thus

$$\frac{\partial \phi}{\partial r} = -2 \frac{\mu}{r^3} \cos \theta, \qquad \frac{\partial \phi}{\partial \theta} = -\frac{\mu}{r^2} \sin \theta$$

and then

$$\frac{\partial \psi}{\partial r} = -\sin \theta \frac{\partial \phi}{\partial \theta} = \frac{\mu}{r^2} \sin^2 \theta$$

$$\frac{\partial \psi}{\partial \theta} = r^2 \sin \theta \frac{\partial \phi}{\partial r} = -\frac{2\mu}{r} \sin \theta \cos \theta$$

Then

$$d\psi = \frac{\partial \psi}{\partial r} dr + \frac{\partial \psi}{\partial \theta} d\theta = \frac{\mu}{r^2} \sin^2 \theta \, dr - \frac{2\mu}{r} \sin \theta \cos \theta \, d\theta$$

must be a perfect differential, as

$$\frac{\partial^2 \psi}{\partial r \, \partial \theta} = \frac{\partial^2 \psi}{\partial \theta \, \partial r}.$$

Integrating the first term with respect to r yields

$$-\frac{\mu}{r} \sin^2 \theta$$

and differentiating with respect to θ yields

$$-\frac{\mu}{r} 2 \sin \theta \cos \theta \, d\theta$$

Subtracting from N, or $(-2\mu/r) \sin \theta \cos \theta \, d\theta$, yields zero. Hence, the solution is

$$\psi = -\frac{\mu}{r} \sin^2 \theta = -\frac{\mu}{r^3} \hat{\omega}^2$$

19. Three-dimensional Sources and Sinks.
A source in three-dimensional flow is a point from which fluid issues at a uniform rate in all directions. It is entirely fictitious, as there is nothing resembling it in nature. That does not, however, reduce its usefulness in obtaining flow patterns. The "strength" of the source m is the rate of flow passing through any surface enclosing the source.

As the flow is outward and is uniform in all directions, the velocity, a distance r from the source, is the strength divided by the area of the sphere through the point with center at the source, or

$$v_r = \frac{m}{4\pi r^2}$$

Since $v_r = -\dfrac{\partial \phi}{\partial r}$ and $v_\theta = 0$, hence $\dfrac{\partial \phi}{\partial \theta} = 0$, and the velocity potential can be found.

$$-\frac{\partial \phi}{\partial r} = \frac{m}{4\pi r^2}$$

and

$$\phi = \frac{m}{4\pi r}$$

A "negative source" is a "sink." Fluid is assumed to flow uniformly into a sink and there disappear.

20. Three-dimensional Doublets. A "doublet," or "double source," is a combination of a source and a sink of equal strength, which are allowed to approach each other in such a manner that the product of their strength and the distance between them remains a constant in the limit. Letting m' be the strength at distance δs apart, then their product $\mu = m' \, \delta s$ remains constant as δs approaches zero. μ is defined as the strength of the doublet. The equations, which will be determined in Sec. 30, have a directional property. The line extending from the sink to the source is the axis of the doublet.

21. Two-dimensional Sources and Sinks. In two-dimensional flow it is customary to consider the flow in planes parallel to the xy-plane, the two planes being considered unit distance apart. A source is a straight line parallel to the z-axis from which fluid is imagined to flow uniformly in all directions *at right angles* to the line. The source appears as a point on the customary two-dimensional flow diagram. The total flow per unit time per unit length of line is called the *strength of the source*. As the flow is in radial lines from the source, the velocity a distance r from the source is $2\pi\mu/2\pi r$, where $2\pi\mu$ indicates its strength. Then

$$-\frac{\partial \phi}{\partial r} = \frac{\mu}{r}, \qquad \frac{\partial \phi}{\partial \theta} = 0$$

and

$$\phi = -\mu \ln r$$

is the velocity potential, where ln indicates the natural logarithm.

A negative source is a sink, into which fluid is imagined to flow uniformly from all directions at right angles to its line.

22. Two-dimensional Doublets. The two-dimensional doublet is defined as the limiting case, as a source and sink of equal strength approach each other so that the product of their strength and the displacement between them remains a constant μ, called the *strength of the doublet.* If a source and sink of constant, finite strength were superposed, there would be no resulting flow, as the sink would absorb all flow from the source. The axis of the doublet is from the sink to the source, *i.e.*,

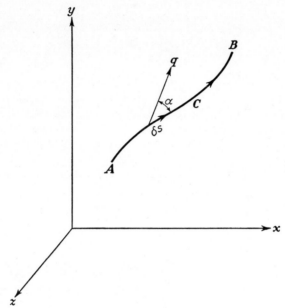

FIG. 14.—Notation used in the definition of a line integral.

the line along which they approached each other. The velocity potential is developed in Sec. 61.

23. The Line Integral. Let $\bar{q}$ be a vector (such as velocity) that is defined throughout a region enclosing the continuous curve C (Fig. 14). The integral

$$\int_C q \cos \alpha \, ds \qquad (27)$$

where α is the angle between the vector and δs, is defined as a line integral of the function $\bar{q}$ over the space curve C. Let u,v,w be the scalar components of $\bar{q}$ in the xyz-directions, and let the direction cosines of $\bar{q}$ and δs be l,m,n and l', m', n', respectively. Then, since $u = lq$, $v = mq$,

$w = nq$ and $\delta x = l' \, \delta s$, $\delta y = m' \, \delta s$, $\delta z = n' \, \delta s$, and

$$\cos \alpha = ll' + mm' + nn'$$

$$\int_C q \cos \alpha \, ds = \int_C (qll' \, ds + qmm' \, ds + qnn' \, ds)$$

$$= \int_C (u \, dx + v \, dy + w \, dz)$$

In general, the value of the line integral between A and B depends upon the particular choice of the curve C. If the equation of the space curve is given in parametric form as

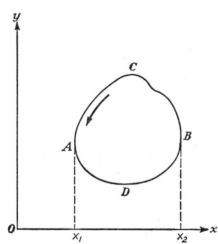

$$x = f_1(t), \qquad y = f_2(t),$$
$$z = f_3(t) \qquad (t_0 \leq t \leq t_1)$$

where f_1, f_2, f_3 possess continuous derivatives in the interval t_0 to t_1, the line integral may be evaluated as a definite integral

$$\int_{t_0}^{t_1} [uf_1'(t) + vf_2'(t) + wf_3'(t)] \, dt$$

where u, v, w are also expressed in terms of t.

FIG. 15.—A closed curve in the xy-plane, with positive direction for traversing the boundary.

Consider the plane closed curve C (Fig. 15) such that no line parallel to the coordinate axes intersects C in more than two points.[1] Let the maximum abscissa of the curve be x_2 and the minimum abscissa be x_1. Let the equation of ACB be given by $y_2 = f_2(x)$ and the equation of ADB be given by $y_1 = f_1(x)$. Then the enclosed area $ADBC$ is given by

$$A = \int_{x_1}^{x_2} y_2 \, dx - \int_{x_1}^{x_2} y_1 \, dx$$
$$= -\int_{x_2}^{x_1} y_2 \, dx - \int_{x_1}^{x_2} y_1 \, dx = -\int_C y \, dx$$

where the last integral is a line integral around the curve taken in the counterclockwise direction. The positive direction is taken as that which permits the area to lie on the left as an observer walks around the periphery on the positive side of the surface (right-handed rule). The restriction on the shape of the area for which this method is applicable may be removed by dividing any actual area into a number of smaller

---

[1] I. S. and E. S. Sokolnikoff, "Higher Mathematics for Engineers and Physicists," 2d ed., pp. 200–201, McGraw-Hill Book Company, Inc., New York, 1941.

areas such that each may be considered as made up of two single-valued curves. In traversing all these peripheries in the positive direction, all internal boundaries will be traversed twice and in opposite directions, so that nothing is contributed by that portion of the line integral. This leaves the formula

$$A = - \int_C y \, dx$$

unrestricted.

Using a similar procedure to that in the foregoing paragraph, a relation that is useful in deriving Stokes' equation is established between a line integral and a surface integral. Let $\dfrac{\partial M(x,y)}{\partial y}$ be a continuous, single-valued function over a region R, bounded by a closed curve C. Then

$$
\begin{aligned}
\int\int_R \frac{\partial M}{\partial y} \, dx \, dy &= \int_{a_1}^{a_2} dx \int_{y_1}^{y_2} \frac{\partial M}{\partial y} \, dy \\
&= \int_{a_1}^{a_2} [M(x,y_2) - M(x,y_1)] \, dx \\
&= - \int_{a_2}^{a_1} M(x,y_2) \, dx - \int_{a_1}^{a_2} M(x,y_1) \, dx \\
&= - \int_C M(x,y) \, dx
\end{aligned}
\tag{28}
$$

where a_1, a_2 are the minimum and maximum x-coordinates on the curve and y_2, y_1 are functions of x representing the upper and lower portions of C, respectively.

24. Stokes' Theorem. Stokes' theorem is used in establishing the relation between circulation and rotation in the following section, and in establishing the Cauchy integral theorem in Chap. VII.

In equation form Stokes' theorem states that

$$
\int (P \, dx + Q \, dy + R \, dz) = \int\int \left(\frac{\partial Q}{\partial x} - \frac{\partial P}{\partial y} \right) dx \, dy \\
+ \int\int \left(\frac{\partial R}{\partial y} - \frac{\partial Q}{\partial z} \right) dy \, dz + \int\int \left(\frac{\partial P}{\partial z} - \frac{\partial R}{\partial x} \right) dz \, dx \tag{29}
$$

where P, Q, R are continuous, differentiable functions of a cartesian system of reference. The line integral on the left-hand side of the equation is carried out along some closed curve in space, and the surface integrals are evaluated over any surface bounded by the curve.

Consider the surface S bounded by the space curve Γ in Fig. 16. Let $z = f(x,y)$ be the equation of the surface S, and let $P(x,y,z)$ be a function that is continuous in the region containing S.   Then

$$\int_{\Gamma} P(x,y,z)\ dx = \int_{C} P[x,y,f(x,y)]\ dx$$

where the second line integral is taken around the projection of Γ in the

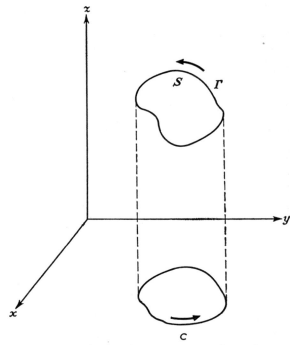

FIG. 16.—Projection of a space curve on the xy-plane.

xy-plane, as shown in Fig. 16.   Differentiating with respect to y,

$$\frac{\partial P[x,y,f(x,y)]}{\partial y} = \frac{\partial P(x,y,z)}{\partial y} + \frac{\partial P(x,y,z)}{\partial z}\frac{\partial z}{\partial y}$$

Now $\dfrac{\partial z}{\partial y} = -\dfrac{m}{n}$ where l,m,n are the direction cosines of the normal δn
to the surface S.   This may be shown as follows: Let $0,m_1,n_1$, be the
direction cosines of an element of length δs in the surface (the element is
at right angles to the x-axis, as $\dfrac{\partial z}{\partial y}$ presupposes $x = $ constant).   Since
δn and δs are also at right angles,

$$l \cdot 0 + m m_1 + n n_1 = 0$$

and

$$-\frac{m}{n} = \frac{n_1}{m_1}$$

Let δy be the displacement of δs in the y-direction and δz be the displacement of δs in the z-direction. Then $\delta y = m_1 \, \delta s$, $\delta z = n_1 \, \delta s$, and

$$\frac{\delta z}{\delta y} = \frac{\partial z}{\partial y} = \frac{n_1}{m_1}$$

hence,

$$\frac{\partial z}{\partial y} = -\frac{m}{n}$$

Applying Eq. (28),

$$\iint_S \frac{\partial P[x,y,f(x,y)]}{\partial y} \, dx \, dy = \iint_S \left[\frac{\partial P(x,y,z)}{\partial y} - \frac{m}{n} \frac{\partial P(x,y,z)}{\partial z} \right] dx \, dy$$

$$= -\int_C P[x,y,f(x,y)] \, dx$$

$$= -\int_\Gamma P(x,y,z) \, dx$$

Rewriting,

$$\int_\Gamma P(x,y,z) \, dx = \iint_S \left[\frac{\partial P(x,y,z)}{\partial z} \, dz \, dx - \frac{\partial P(x,y,z)}{\partial y} \, dx \, dy \right]$$

since

$$\delta S = \frac{\delta x \, \delta y}{n} \qquad \text{and} \qquad m \, \delta S = \delta x \, \delta z$$

In a similar manner it may be shown that

$$\int_\Gamma Q(x,y,z) \, dy = \iint_S \left(\frac{\partial Q}{\partial x} \, dx \, dy - \frac{\partial Q}{\partial z} \, dy \, dz \right)$$

and

$$\int_\Gamma R(x,y,z) \, dz = \iint_S \left(\frac{\partial R}{\partial y} \, dy \, dz - \frac{\partial R}{\partial x} \, dz \, dx \right)$$

Adding the last three equations, Stokes' theorem in its most general form [Eq. (29)] is obtained.

25. Circulation. The line integral of the velocity vector taken around a closed curve enclosing a surface S within a fluid region is said to be the *circulation* Γ. Referring to Sec. 23, this is

$$\Gamma = \int_C (u \, dx + v \, dy + w \, dz) \tag{30}$$

Using Stokes' theorem, where P, Q, R are replaced by u, v, w, which are also scalar functions of space,

$$\Gamma = \int \int \left(\frac{\partial v}{\partial x} - \frac{\partial u}{\partial y} \right) dx \, dy + \int \int \left(\frac{\partial w}{\partial y} - \frac{\partial v}{\partial z} \right) dy \, dz$$

$$+ \int \int \left(\frac{\partial u}{\partial z} - \frac{\partial w}{\partial x} \right) dz \, dx \quad (31)$$

The first integrand is $2\omega_z$, the second integrand $2\omega_x$, and the third integrand $2\omega_y$, where ω_x, ω_y, ω_z are the rotation components derived in Sec. 9. Rewriting Eq. (31),

$$\Gamma = 2\int (l\omega_x + m\omega_y + n\omega_z) \, dS \qquad (32)$$

where l, m, n are the direction cosines of the normal to δS, it is obvious that the circulation about any infinitesimal area δS is equal to twice the product of the area element and the component of the rotation vector normal to the surface,

$$\delta\Gamma = 2\omega_n \, \delta S \qquad (33)$$

As any surface may be divided into surface elements δS, in words: The circulation about any closed curve is equal to twice the surface integral of the normal rotation component over the surface enclosed by the curve. A restriction on this statement is that no singular points occur on the surface or boundary; i.e., ω_n is continuous and defined at all points on the surface.

When the flow is irrotational, the right-hand side of Eq. (31) is zero; hence, there is no circulation about any closed curve in the flow not enclosing a singular point. Conversely, if the circulation about every closed curve in a fluid is zero, when the curve contains no singular points, the flow is irrotational. This is apparent from Eq. (33), as the closed curve may be in any position and be arbitrarily small. It is, therefore, necessary that the rotation be everywhere zero for the circulation to be zero. Circulation has the dimensions L^2T^{-1}.

26. Vortices. An example of two-dimensional flow with circulation is given by the velocity potential

$$\phi = -\mu\theta$$

where μ is a constant and θ is the angle used in plane polar coordinates, measured positive in a counterclockwise direction from the x-axis. The velocity in a radial direction is everywhere zero, as $-\dfrac{\partial \phi}{\partial r} = 0$. The tangential velocity at any point r distance from the origin is

$$v_\theta = -\frac{\partial \phi}{r \, \partial \theta} = \frac{\mu}{r}$$

as $r\,\delta\theta$ is the element of length in the tangential direction. Since the velocity is constant about any circle of radius r, the circulation, from

$$\Gamma = \int_C q \cos \alpha \, ds$$

is

$$\Gamma = q \int ds = qs = \frac{\mu}{r} 2\pi r = 2\pi\mu$$

as $\alpha = 0$. At the origin the velocity becomes infinite $(r = 0)$, which makes it a singular point. As the velocity is defined for all other points in the plane, the origin is the only singular point.

The flow described above is vortex flow, with a vortex of strength $2\pi\mu$ at the origin. The flow is irrotational at all points except the origin. The circulation is zero about every closed curve in the plane that excludes the origin and is $2\pi\mu$ for all curves that contain the origin.

27. Summary. In this chapter the principle of work and energy was established for a nonviscous fluid, a formula for kinetic energy in terms of velocity potential was derived, and several uniqueness theorems were proved as a direct consequence. The stream functions in two- and three-dimensional flow were defined, as well as the basic flows set up by sources, doublets, and vortices. The relations between rotation and circulation were obtained. Sufficient theory has now been made available so that many of the classical three-dimensional flow cases may be studied.

Exercises

1. Let v_r, v_θ be the velocity components in the radial and tangential directions in plane polar coordinates.

(a) Express v_r, v_θ in terms of the velocity potential $\phi(r, \theta)$.

(b) Express v_r, v_θ in terms of the stream function $\psi(r, \theta)$.

(c) Find the differential equations relating ϕ to ψ.

Ans.   (a) $v_r = -\dfrac{\partial \phi}{\partial r}$, $v_\theta = -\dfrac{\partial \phi}{r\,\partial \theta}$; (b) $v_r = -\dfrac{\partial \psi}{r\,\partial \theta}$, $v_\theta = \dfrac{\partial \psi}{\partial r}$; (c) $\dfrac{\partial \phi}{\partial r} = \dfrac{1}{r}\dfrac{\partial \psi}{\partial \theta}$,

$$\frac{\partial \phi}{\partial \theta} = -r \frac{\partial \psi}{\partial r}.$$

2. The velocity potential for a three-dimensional source of strength m is

$$\phi = \frac{m}{4\pi r}$$

from Sec. 19.   Find the Stokes' stream function for this flow.

$$\textit{Ans.} \quad \psi = \frac{m}{4\pi} \cos \theta.$$

3. The velocity potential for a two-dimensional source of strength $2\pi\mu$ is

$$\phi = -\mu \ln r$$

from Sec. 21.   Find the stream function for this flow. $\textit{Ans.}$ $\psi = -\mu\theta$.

4. Evaluate the line integral of the velocity $u = y$, $v = z$, $w = x$ over the path $x = t$, $y = t^2 - 1$, $z = 2t$ $(0 < t < 2)$; velocities are in feet per second, lengths in feet, and time t in seconds.   *Ans.*   15.33 ft² per sec.

5. Evaluate the line integral of the velocity over the closed path

$$(x - 1)^2 + (y - 6)^2 = 4,$$

when $u = 3x + y$, $v = 2x - 3y$; x and y are in feet, v in feet per second.

Ans. 4π ft² per sec.

6. Find the stream function for a vortex of strength $2\pi\mu$ at the origin.

Ans. $\psi = \mu \ln r$.

7. Express the velocity potential and stream function of Exercise 6 in cartesian coordinates.

8. Find the velocity potential for a source of strength 5 ft³ per sec per foot of width at $x = 3$, $y = 4$.

9. Find the circulation about the square enclosed by the lines $x = \pm 1$, $y = \pm 1$ for the two-dimensional flow given by $u = x + y$, $v = x^2 - y$.

Ans. -4 ft² per sec.

10. Show that $\phi = xy$ satisfies the Laplace equation for two-dimensional flow, $\nabla^2 \phi = \dfrac{\partial^2 \phi}{\partial x^2} + \dfrac{\partial^2 \phi}{\partial y^2} = 0$.   Determine the stream function.   Show that the equipotential lines and the streamlines are orthogonal.   Are there any exceptions?

11. Let $\psi = \frac{1}{2} U r^2 \sin^2 \theta$ for three-dimensional flow, axial symmetry, where U is a constant.   Find the discharge between stream surfaces through $r = 10$, $\theta = \pi/6$, and $r = 1$, $\theta = \pi/4$.   Draw the streamlines $\psi = U, 2U, 3U$.

12. Express the stream function of Exercise 11 in cartesian coordinates and obtain the potential function.   What is the nature of the flow?

13. If a velocity distribution satisfies continuity everywhere and has zero rotation at several particular points, can it be inferred that the flow is irrotational except for a finite number of singular points in space?

CHAPTER IV

THREE-DIMENSIONAL FLOW EXAMPLES

28. Source in a Uniform Stream. The velocity potential for a source of strength m located at the origin was worked out in Sec. 19 to be

$$\phi = \frac{m}{4\pi r} \tag{1}$$

where r is the distance from the origin. The radial velocity is

$$v_r = -\frac{\partial \phi}{\partial r} = \frac{m}{4\pi r^2}$$

which, when multiplied by the surface area of the sphere concentric with it $4\pi r^2$, gives the strength m. Since the flow from the source is symmetrical with respect to the x-axis, the stream function is defined. For spherical polar coordinates, from Eqs. (22), Sec. 17,

$$\frac{\partial \psi}{\partial r} = -\sin \theta \frac{\partial \phi}{\partial \theta}, \qquad \frac{\partial \psi}{\partial \theta} = r^2 \sin \theta \frac{\partial \phi}{\partial r}$$

Using Eq. (1),

$$\frac{\partial \psi}{\partial r} = 0, \qquad \frac{\partial \psi}{\partial \theta} = -\frac{m}{4\pi} \sin \theta$$

Integrating, as in Sec. 18,

$$\psi = \frac{m}{4\pi} \cos \theta \tag{2}$$

is the stream function for a source at the origin. Equipotential lines and streamlines are shown in Fig. 17 for constant increments of ϕ and ψ.

A uniform stream of fluid having a velocity U in the negative x-direction throughout space is given by

$$-\frac{\partial \phi}{\partial x} = -U, \qquad \frac{\partial \phi}{\partial \hat{\omega}} = 0$$

Integrating,

$$\phi = Ux = Ur \cos \theta \tag{3}$$

The stream function is found in the same manner as above to be

$$\psi = \frac{U}{2} \hat{\omega}^2 = \frac{Ur^2}{2} \sin^2 \theta \tag{4}$$

The flow network is shown in Fig. 18.

Combining the uniform flow and the source flow, which may be accomplished by adding the two velocity potentials and the two stream functions, gives

$$\left.\begin{array}{l} \phi = \dfrac{m}{4\pi r} + Ur\cos\theta \\[3mm] \psi = \dfrac{m}{4\pi}\cos\theta + \dfrac{Ur^2}{2}\sin^2\theta \end{array}\right\} \tag{5}$$

The resulting flow is everywhere the same as if the separate velocity vectors were added for each point in space.

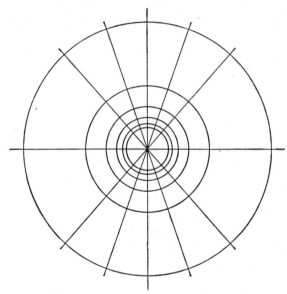

Fig. 17.—Streamlines and equipotential lines for a source.

A stagnation point is a point in the fluid where the velocity is zero. The conditions for stagnation point, where spherical polar coordinates are used and when the flow has axial symmetry, are

$$v_r = -\frac{\partial\phi}{\partial r} = 0, \qquad v_\theta = -\frac{1}{r}\frac{\partial\phi}{\partial\theta} = 0$$

Use of these expressions with Eqs. (5) gives

$$\frac{m}{4\pi r^2} - U\cos\theta = 0, \qquad U\sin\theta = 0$$

which are satisfied by only one point in space, *viz.*,

$$\theta = 0, \qquad r = \sqrt{\frac{m}{4\pi U}}$$

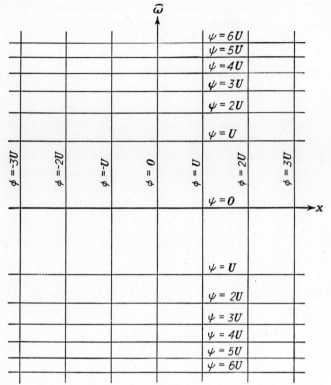

Fig. 18.—Streamlines and equipotential lines for uniform flow in the negative x-direction.

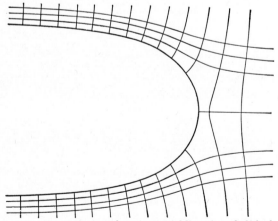

Fig. 19.—Streamlines and equipotential lines for a half body.

Substituting this point back into the stream function gives $\psi = m/4\pi$, which is the stream surface through the stagnation point. The equation of this surface is found from Eqs. (5):

$$\cos \theta + \frac{2\pi U}{m} r^2 \sin^2 \theta = 1 \qquad (6)$$

The flow under consideration is steady, as the velocity potential does not change with the time. Therefore, any stream surface satisfies the conditions for a boundary: The velocity component normal to the stream surface in steady flow is always zero. Since stream surfaces through stagnation points usually split the flow, they are frequently the most interesting possible boundary. This stream surface is plotted in Fig. 19. Substituting $\hat\omega = r \sin \theta$ in Eq. (6), the distance of a point (r,θ) from the x-axis is given by

$$\hat\omega^2 = \frac{m}{2\pi U} (1 - \cos \theta)$$

which shows that $\hat\omega$ has a maximum value $\sqrt{m/\pi U}$ as θ approaches π, i.e., as r approaches infinity. Hence, $\hat\omega = \sqrt{m/\pi U}$ is an asymptotic surface to the dividing stream surface. Equation (6) may be expressed in the form

$$r = \frac{1}{2} \sqrt{\frac{m}{\pi U}} \sec \frac{\theta}{2} \qquad (7)$$

from which the surface is easily plotted. Such a figure of revolution is called a *half body*, as it extends to negative infinity, surrounding the negative x-axis.

The pressure intensity at any point, *i.e.*, the dynamic pressure from Eq. (22), Sec. 10, is

$$p = \frac{\rho}{2} (U^2 - q^2)$$

where the dynamic pressure at infinity is taken as zero. q is the total velocity at any point. Evaluating q from Eqs. (5),

$$q^2 = \left(\frac{\partial \phi}{\partial r}\right)^2 + \frac{1}{r^2} \left(\frac{\partial \phi}{\partial \theta}\right)^2 = U^2 + \frac{m^2}{16\pi^2 r^4} - \frac{mU \cos \theta}{2\pi r^2}$$

and

$$p = \frac{\rho}{2} U^2 \left(\frac{m \cos \theta}{2\pi r^2 U} - \frac{m^2}{16\pi^2 r^4 U^2}\right) \qquad (8)$$

from which the pressure can be found for any point except the origin, which is a singular point. Substituting Eq. (7) into Eq. (8), the pressure intensity is given in terms of r for any point on the half body; thus

$$p = \frac{\rho}{2} U^2 \left(\frac{3m^2}{16\pi^2 r^4 U^2} - \frac{m}{2\pi r^2 U} \right) \qquad (9)$$

This shows that the dynamic pressure approaches zero as r increases downstream along the body.

29. Source and Sink of Equal Strength in a Uniform Stream. Rankine Bodies. A source of strength m, located at (a,o), has the velocity potential at any point P given by

$$\phi_1 = \frac{m}{4\pi r_1}$$

where r_1 is the distance from (a,o) to P, as shown in Fig. 20. Similarly, the potential function for a sink of strength m at $(-a,o)$ is

$$\phi_2 = - \frac{m}{4\pi r_2}$$

Since both ϕ_1 and ϕ_2 satisfy the Laplace equation, their sum will also be a solution,

$$\phi = \frac{m}{4\pi} \left(\frac{1}{r_1} - \frac{1}{r_2} \right) \qquad (10)$$

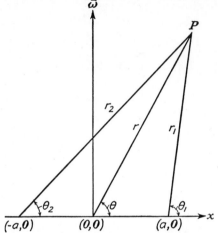

FIG. 20.—Auxiliary coordinate systems used for Rankine body.

Because r_1, r_2 are measured from different points, this expression must be handled differently from the usual algebraic equation.

The stream functions for the source and sink may also be added to give the stream function for the combined flow

$$\psi = \frac{m}{4\pi} (\cos \theta_1 - \cos \theta_2) \qquad (11)$$

The stream surfaces and equipotential surfaces take the form shown in Fig. 21, which is plotted from Eqs. (10) and (11) by taking constant values of ϕ and ψ.

Superposing a uniform flow of velocity U in the negative x-direction, $\phi = Ux$, $\psi = \frac{1}{2}U\bar{\omega}^2$, the potential and stream functions for source and sink of equal strength in a uniform flow (in direction of source to sink) are

$$\phi = Ux + \frac{m}{4\pi}\left(\frac{1}{r_1} - \frac{1}{r_2}\right)$$

$$= Ux + \frac{m}{4\pi}\left[\frac{1}{\sqrt{(x-a)^2 + \hat{\omega}^2}} - \frac{1}{\sqrt{(x+a)^2 + \hat{\omega}^2}}\right] \quad (12)$$

$$\psi = \tfrac{1}{2}Ur^2 \sin^2\theta + \frac{m}{4\pi}(\cos\theta_1 - \cos\theta_2) \quad (13)$$

As any stream surface may be taken as a solid boundary in steady

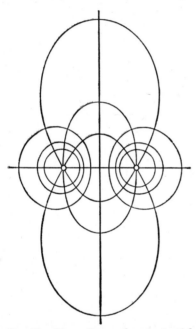

flow, the location of a closed surface for this flow case will represent flow of a uniform stream around a body. Examining the stream function, for $x > a$ and $\theta_1 = \theta_2 = \theta = 0$, $\psi = 0$. For $x < -a$ and $\theta_1 = \theta_2 = \theta = \pi$, $\psi = 0$. Therefore, $\psi = 0$ must be the dividing streamline, since the x-axis is the axis of symmetry. The equation of the dividing streamline is, from Eq. (13),

$$\hat{\omega}^2 + \frac{m}{2\pi U}(\cos\theta_1 - \cos\theta_2) = 0 \quad (14)$$

where $\hat{\omega} = r\sin\theta$ is the distance of a point on the dividing stream surface from the x-axis. Since $\cos\theta_1$ and $\cos\theta_2$ are never greater than unity, $\hat{\omega}$ cannot exceed $\sqrt{\dfrac{m}{\pi U}}$, which shows

Fig. 21.—Streamlines and equipotential lines for source and sink of equal strength.

that the surface is closed, and hence can be replaced by a solid body of exactly the same shape. By changing the signs of m and U the flow is reversed and the body should change end for end. From Eq. (14) it is seen that the equation is unaltered; hence, the body has symmetry with respect to the plane $x = 0$. It is necessarily a body of revolution because of axial symmetry of the equations.

To locate the stagnation points C, D (Fig. 22), which must be on the x-axis, it is known that the velocity is along the x-axis (it is a streamline). From Eq. (12) the velocity potential ϕ_x for points on the x-axis is given by

$$\phi_x = \frac{ma}{2\pi} \frac{1}{x^2 - a^2} + Ux$$

since

$$r_1 = x - a, \qquad r_2 = x + a$$

Differentiating with respect to x and setting the result equal to zero,

$$\frac{\partial \phi_x}{\partial x} = U - \frac{max_0}{\pi (x_0{}^2 - a^2)^{\frac{3}{2}}} = 0 \tag{15}$$

where x_0 is the x-coordinate of the stagnation point.   This gives the point

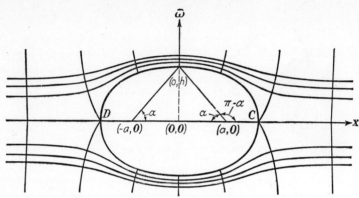

Fig. 22.—Rankine body.

$C(x_0,0)$ (a trial solution).   The half breadth h is determined as follows:
From Fig. 22

$$\theta_1 = \pi - \alpha, \qquad \theta_2 = \alpha$$

where

$$\cos \alpha = \frac{a}{\sqrt{h^2 + a^2}}$$

Substituting into Eq. (14),

$$h^2 = \frac{m}{\pi U} \frac{a}{\sqrt{h^2 + a^2}} \tag{16}$$

from which h may be determined (also by trial solution).
   Eliminating m/U between Eqs. (15) and (16),

$$\frac{m}{U} = \frac{(x_0{}^2 - a^2)^2}{x_0} \frac{\pi}{a} = \frac{\pi}{a} h^2 \sqrt{h^2 + a^2}$$

the value of a may be obtained for a predetermined body (x_0, h, specified).
Hence, U can be given any positive value and the pressure and velocity
distribution can be determined.

In determining the velocity at points throughout the region it is convenient to find the velocity at each point due to each component of the flow, i.e., due to the source, the sink, and the uniform flow, separately, and add the components graphically or by $\hat{\omega}$- and x-components.

Bodies obtained from source-sink combinations with uniform flow are called *Rankine bodies*.

30. Equation for Three-dimensional Doublet. In Sec. 20 the doublet was defined as the limiting case of a source and sink of equal strength such that the product of the distance between them and their strength remains constant as they approach each other. The limiting process is applied to Eq. (10), where $2a$ approaches zero as $2a$ times the strength m is held constant and equal to $4\pi\mu$. μ is defined as the strength of the doublet.

Referring to Fig. 20, by the law of sines,

$$\frac{r_1}{\sin \theta_2} = \frac{r_2}{\sin \theta_1} = \frac{2a}{\sin (\theta_1 - \theta_2)}$$
$$= \frac{2a}{2 \sin \frac{1}{2}(\theta_1 - \theta_2) \cos \frac{1}{2}(\theta_1 - \theta_2)}$$

as the angle between r_2 and r_1 at P is $\theta_1 - \theta_2$. Solving for $r_2 - r_1$,

$$r_2 - r_1 = \frac{a(\sin \theta_1 - \sin \theta_2)}{\sin \frac{1}{2}(\theta_1 - \theta_2) \cos \frac{1}{2}(\theta_1 - \theta_2)}$$
$$= \frac{2a \cos \frac{1}{2}(\theta_1 + \theta_2)}{\cos \frac{1}{2}(\theta_1 - \theta_2)}$$

From Eq. (10)

$$\phi = \frac{m}{4\pi} \frac{r_2 - r_1}{r_1 r_2} = \frac{2am \cos \frac{1}{2}(\theta_1 + \theta_2)}{4\pi r_1 r_2 \cos \frac{1}{2}(\theta_1 - \theta_2)}$$
$$= \frac{\mu}{r_1 r_2} \frac{\cos \frac{1}{2}(\theta_1 + \theta_2)}{\cos \frac{1}{2}(\theta_1 - \theta_2)}$$

In the limit as a approaches zero, $\theta_2 = \theta_1 = \theta$, $r_2 = r_1 = r$, and

$$\phi = \frac{\mu}{r^2} \cos \theta \tag{17}$$

which is the velocity potential for a doublet[1] at the origin with axis in the positive x-direction. Equation (17) may be converted into the stream function by Eqs. (22), Sec. 17. The stream function is

$$\psi = -\frac{\mu\hat{\omega}^2}{r^3} = -\frac{\mu \sin^2 \theta}{r} \tag{18}$$

[1] L. M. Milne-Thompson, "Theoretical Hydrodynamics," p. 414, Macmillan & Co., Ltd., London, 1938.

The stream function may also be obtained by applying the limiting process to Eq. (11). Streamlines and equipotential lines for the doublet are drawn in Fig. 23.

The combination of a doublet and uniform flow in the direction of the negative axis should give some form of closed body, as it is the limiting case for a Rankine body. In fact, it is the case of flow around a sphere that is treated in Sec. 34.

31. Finite Line Source. A line source is a line over which infinitesimal point sources are continuously distributed. The strength m is defined as the flow out from the line per unit length. For an element of line $d\xi$, the strength of elemental source is $m\,d\xi$. From Eq. (2) the stream function at some point P for the elemental source is

Fig. 23.—Streamlines and equipotential lines for a three-dimensional doublet.

$$\delta\psi = \frac{m}{4\pi}\,\delta\xi\,\cos\alpha$$

where α is the angle the radial line from the source makes with the x-axis, as shown in Fig. 24.

For the stream function to be defined there must be axial symmetry; hence, the line source must be straight. It is convenient to consider the line source as extending from the origin along the positive x-axis to $(a,0)$. As the stream functions may be superposed, the value of ψ at P due to the line source from O to a is

$$\psi = \frac{1}{4\pi}\int_{o}^{a} m \cos\alpha\,d\xi$$

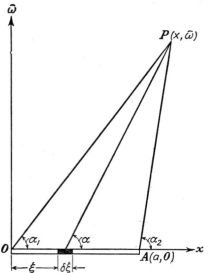

Fig. 24.—Line source along the x-axis.

To perform the integration, m and α must be expressed in terms of ξ. Restricting the problem to constant m, this is easily effected, as

$$x - \xi = \hat{\omega} \cot\alpha, \qquad d\xi = \hat{\omega} \csc^2\alpha\,d\alpha$$

and

$$\psi = \frac{m\hat{\omega}}{4\pi} \int_{\alpha_1}^{\alpha_2} \frac{\cos \alpha}{\sin^2 \alpha}\, d\alpha = \frac{m\hat{\omega}}{4\pi}\left(\frac{1}{\sin \alpha_1} - \frac{1}{\sin \alpha_2}\right) = \frac{m}{4\pi}\,(\overline{PO} - \overline{PA})$$

The streamlines in a plane through the axis of symmetry are hyperbolas with foci at O and A.

A line sink is a negative line source.   Combinations of uniform flow with point sources and sinks as well as line sources and sinks permit velocity and pressure distributions to be determined for a variety of bodies.[1] Equipotential lines and streamlines are drawn in Fig. 25.

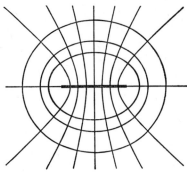

32. The Laplace Equation in Spherical Coordinates. The intersection of three orthogonal surfaces defines a point in spherical polar coordinates. These surfaces are a sphere, a circular cone, and a plane. The sphere is concentric with the origin and has the equation

Fig. 25.—Streamlines and equipotential lines for finite line source.

$$x^2 + y^2 + z^2 = r^2.$$

The cone has its vertex at the origin and axis coincident with the x-axis. The angle made by an element of the cone and the x-axis is θ, known as the *polar angle*. The plane is through the x-axis, and its orientation is given by the angle ω it makes with the y-axis, as shown in Fig. 26.   The following relations may be established from the figure:

$$x = r \cos \theta, \qquad y = r \sin \theta \cos \omega, \qquad z = r \sin \theta \sin \omega \qquad (19)$$

The elements of length δs in the $r\theta\omega$-directions are

$$\delta s_r = \delta r, \qquad \delta s_\theta = r\,\delta\theta, \qquad \delta s_\omega = r \sin \theta\,\delta\omega$$

and form a right-handed system in the order given. The element of volume is their product

$$\delta V = r^2 \sin \theta\,\delta r\,\delta\theta\,\delta\omega$$

The Laplace equation is now derived by writing the condition that the net rate of flow (incompressible fluid) out of the elemental volume is zero.   Let the center of the volume element be at (r,θ,ω), and consider the rate of flow δQ through the face at (r,θ,ω) normal to δr as in Fig. 27.

[1] For additional information and references see R. H. Smith, Aerodynamic Theory and Test of Strut Forms, *NACA Reports* 311 and 335, 1929.

In terms of the velocity potential it is

$$\delta Q = -\frac{\partial \phi}{\partial r} r \, \delta\theta \, r \sin\theta \, \delta\omega$$

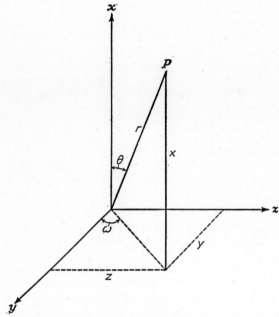

FIG. 26.—Notation used for spherical polar coordinates.

Then the flow out of the parallel face at $r + \dfrac{\delta r}{2}$ minus the flow into the

parallel face at $r - \dfrac{\delta r}{2}$ is

$$\left[\delta Q + \frac{\partial}{\partial r}(\delta Q)\frac{\delta r}{2} \right] - \left[\delta Q - \frac{\partial}{\partial r}(\delta Q)\frac{\delta r}{2} \right]$$

or substituting the value of δQ,

$$\frac{\partial}{\partial r}\left(-\frac{\partial\phi}{\partial r} r \, \delta\theta \, r \sin\theta \, \delta\omega \right)\delta r$$

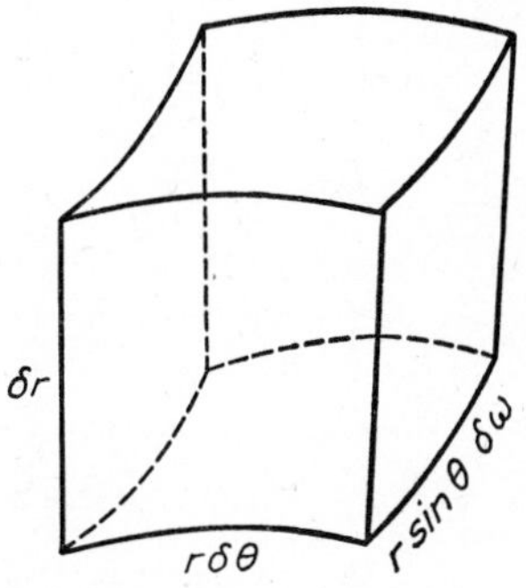

For the pair of faces normal to the $r\,\delta\theta$ direction the excess rate of flow out is similarly

FIG. 27.—Volume element in spherical polar coordinates.

$$\frac{\partial}{r\,\partial\theta}\left(-\frac{\partial\phi}{r\,\partial\theta}\,\delta r\, r\sin\theta\,\delta\omega \right)r\,\delta\theta$$

and for the remaining pair of faces

$$\frac{\partial}{r \sin \theta \, \partial \omega} \left(-\frac{\partial \phi}{r \sin \theta \, \partial \omega} \, \delta r \, r \, \delta \theta \right) r \sin \theta \, \delta \omega$$

Since the net excess rate from all six faces must be zero, the Laplace equation is obtained by adding the three expressions and simplifying; thus

$$\frac{\partial}{\partial r} \left(r^2 \frac{\partial \phi}{\partial r} \right) + \frac{1}{\sin \theta} \frac{\partial}{\partial \theta} \left(\sin \theta \frac{\partial \phi}{\partial \theta} \right) + \frac{1}{\sin^2 \theta} \frac{\partial^2 \phi}{\partial \omega^2} = 0 \qquad (20)$$

For symmetry with respect to the x-axis the last term becomes zero. Equation (20) may also be obtained directly from Eq. (23), Sec. 11, by change of variables, using Eqs. (19).

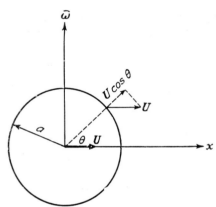

Fig. 28.—Sphere translating in the positive x-direction.

33. Translation of a Sphere in an Infinite Fluid. Virtual Mass. The velocity potential for a solid moving through an infinite fluid otherwise at rest must satisfy the following conditions:[1]

1. The Laplace equation, $\nabla^2 \phi = 0$ everywhere except singular points.

2. The fluid must remain at rest at infinity; hence, the space derivatives of ϕ must vanish at infinity.

3. The boundary conditions at the surface of the solid must be satisfied; *i.e.*, $\dfrac{DF}{Dt} = 0$ [Eq. (15), Sec. 8].

For a sphere of radius a with center at the origin moving with velocity U in the positive x-direction, the velocity of the surface normal to itself is $U \cos \theta$, from Fig. 28. The fluid velocity normal to the surface is $-\dfrac{\partial \phi}{\partial r}$; hence the boundary condition is

$$-\frac{\partial \phi}{\partial r} = U \cos \theta$$

---

[1] G. G. Stokes, "Mathematical and Physical Papers," Vol. 1, pp. 38–43, Cambridge University Press, London, 1880.

for $r = a$. This may be obtained from the general boundary equation in the following manner:

$$F = (x - Ut)^2 + \hat{\omega}^2 - a^2 = 0$$

is the equation of the moving surface, such that when $t = 0$, the center is at the origin. Then

$$\frac{DF}{Dt} = u \frac{\partial F}{\partial x} + v' \frac{\partial F}{\partial \hat{\omega}} + \frac{\partial F}{\partial t} = 0$$

and since $u = -\dfrac{\partial \phi}{\partial x}$, $v' = -\dfrac{\partial \phi}{\partial \hat{\omega}}$ for $t = 0$, the condition becomes

$$-x \frac{\partial \phi}{\partial x} - \hat{\omega} \frac{\partial \phi}{\partial \hat{\omega}} - Ux = 0$$

for $r = a$. Dividing the equation through by a, as $x/a = \cos \theta$, $\hat{\omega}/a = \sin \theta$, for $r = a$, the first two terms give the velocity in the radial direction $-\dfrac{\partial \phi}{\partial r}$, and the desired boundary condition is procured.

The fluid flow about a sphere moving along the x-axis will have axial symmetry with respect to the x-axis. Hence, the last term of Eq. (20) drops out, leaving

$$\frac{\partial}{\partial r} \left(r^2 \frac{\partial \phi}{\partial r} \right) + \frac{1}{\sin \theta} \frac{\partial}{\partial \theta} \left(\sin \theta \frac{\partial \phi}{\partial \theta} \right) = 0 \qquad (21)$$

to be satisfied by the velocity potential. The velocity potential for the doublet [Eq. (17)]

$$\phi = \frac{\mu \cos \theta}{r^2}$$

satisfies Eq. (21) for any constant value of μ. Substituting it into the boundary condition

$$-\frac{\partial \phi}{\partial r} = \frac{2\mu}{r^3} \cos \theta = U \cos \theta$$

which is satisfied for $r = a$ if $\mu = Ua^3/2$. It may also be noted that the velocity components, $-\dfrac{\partial \phi}{\partial r}$ and $-\dfrac{1}{r} \dfrac{\partial \phi}{\partial \theta}$, are zero at infinity. Therefore,

$$\phi = \frac{Ua^3}{2r^2} \cos \theta \qquad (22)$$

satisfies all the conditions for translation of a sphere in an infinite fluid. This case is one of unsteady flow, solved for the instant when the center of the sphere is at the origin. Because this equation has been specialized for a particular instant, the pressure distribution cannot be found from it by use of Eq. (21), Sec. 10. Streamlines and equipotential lines for the sphere are shown in Fig. 29.

The stream function for this flow case is

$$\psi = -\frac{Ua^3}{2r}\sin^2\theta \qquad (23)$$

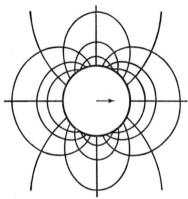

The kinetic energy of the fluid is obtained from Eq. (11), Sec. 14:

$$T = -\frac{\rho}{2}\int \phi \frac{\partial\phi}{\partial n}\,dS$$

$$= \frac{\rho U^2 a^6}{4r^5}\int \cos^2\theta\,dS$$

where the integration is carried out over the surface of the sphere $r = a$. The surface element may be taken as in Fig. 30. It is

$$dS = 2\pi a\sin\theta\,a\,d\theta$$

FIG. 29.—Streamlines and equipotential lines for a sphere moving through fluid.

Substitution into the kinetic energy equation, for $r = a$, yields

$$T = \frac{\rho\pi U^2 a^3}{2}\int_0^\pi \cos^2\theta\sin\theta\,d\theta = \frac{1}{2}\left(\frac{2}{3}\pi a^3\rho\right)U^2$$

Written in this form the kinetic energy of all the fluid is equal to the kinetic energy of one-half the displaced mass of the fluid if it were moving with the velocity U. Letting M' represent $\frac{2}{3}\pi a^3\rho$ or one-half the mass of fluid displaced by the sphere, the kinetic energy of sphere and fluid together is

$$T' = \tfrac{1}{2}(M + M')U^2$$

provided the sphere is not in rotation. M is the mass of the sphere. Since the work done per unit time on the sphere and fluid must equal the time rate of increase of kinetic energy,

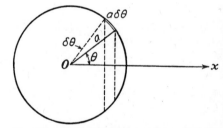

FIG. 30.—Area element for surface of sphere.

$$FU = \frac{dT'}{dt} = MU\frac{dU}{dt} + M'U\frac{dU}{dt}$$

where F is the resultant force acting on the sphere (buoyant, gravity, or other external force). From this equation it is evident that the sphere experiences a resistance $M' \dfrac{dU}{dt}$ due to presence of the fluid. This resistance is zero for constant velocity U. The term M' is known as the "virtual mass" of the sphere. The fact that the fluid offers no resistance to the passage of a solid through it at constant speed is peculiar to ideal fluid theory only. Any body passing through a real fluid experiences a resisting force, called a *drag*, which depends upon the form of the body and its surface roughness.

34. Steady Flow of an Infinite Fluid around a Sphere. The unsteady flow case in the preceding section may be converted into a steady flow case by superposing upon the flow a uniform stream of magnitude U in the negative x-direction. To prove this, add $\phi = Ux = Ur \cos \theta$ to the potential function [Eq. (22)]; thus

$$\phi = \frac{Ua^3}{2r^2} \cos \theta + Ur \cos \theta \tag{24}$$

The stream function corresponding to this is

$$\psi = -\frac{Ua^3}{2r} \sin^2 \theta + \frac{Ur^2}{2} \sin^2 \theta \tag{25}$$

Then from Eq. (25), $\psi = 0$ when $\theta = 0$ and when $r = a$. Hence, the stream surface $\psi = 0$ is the sphere $r = a$, which may be taken as a solid, fixed boundary. Streamlines and equipotential lines are shown in Fig. 31. Perhaps mention should be made that the equations give a flow pattern for the interior portion of the sphere as well. No fluid passes through the surface of the sphere, however.

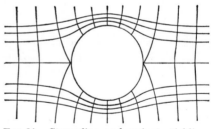

FIG. 31.—Streamlines and equipotential lines for uniform flow about a sphere at rest.

The velocity at any point on the surface of the sphere is

$$-\frac{1}{r} \frac{\partial \phi}{\partial \theta}\bigg]_{r=a} = q = \tfrac{3}{2} U \sin \theta$$

The stagnation points are at $\theta = 0$, $\theta = \pi$. The maximum velocity $\tfrac{3}{2} U$ occurs at $\theta = \pi/2$. The dynamic pressure distribution over the surface of the sphere is

$$p = \frac{\rho U^2}{2} (1 - \tfrac{9}{4} \sin^2 \theta)$$

for dynamic pressure of zero at infinity.

35. Fluid Motion Due to the Translation of a Sphere within a Fixed Concentric Spherical Shell. This is an unsteady fluid flow case in which the resulting flow pattern is valid only for the instant the sphere is concentric with the shell. The shell has a radius b, and its center is fixed at the origin. The sphere of radius a has a velocity U in the positive x-direction and has its center at the origin at the instant the analysis of flow conditions is made.[1]

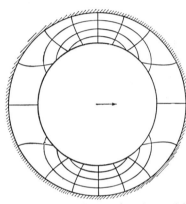

Fig. 32.—Streamlines and equipotential lines for motion of a sphere in a fixed concentric spherical shell.

The boundary condition at the surface of the sphere, $r = a$, is

$$-\frac{\partial \phi}{\partial r} = U \cos \theta$$

as in Sec. 33. At the fixed surface $r = b$

$$\frac{\partial \phi}{\partial r} = 0$$

must be satisfied.

A potential function is now assumed and shown to satisfy the necessary conditions:

$$\phi = \left(Ar + \frac{B}{r^2}\right) \cos \theta$$

where A, B are arbitrary. Substitution in Eq. (21) shows that $\nabla^2\phi = 0$ is satisfied for any values of A, B. Substituting the boundary conditions,

$$U \cos \theta = -\left(A - \frac{2B}{a^3}\right) \cos \theta, \qquad \left(A - \frac{2B}{b^3}\right) \cos \theta = 0$$

from which

$$A = \frac{Ua^3}{b^3 - a^3}, \qquad B = \frac{U}{2} \frac{a^3 b^3}{b^3 - a^3}$$

Hence,

$$\phi = \frac{a^3 U}{b^3 - a^3}\left(r + \frac{b^3}{2r^2}\right) \cos \theta \tag{26}$$

is the velocity potential. It is not necessary to consider conditions at infinity, since the fluid is bounded externally. The stream function

[1] G. G. Stokes, "Mathematical and Physical Papers," Vol. 1, pp. 38–43, Cambridge University Press, London, 1880.

corresponding to Eq. (26) is

$$\psi = \frac{a^3 U}{2(b^3 - a^3)} \left(r^2 - \frac{b^3}{r} \right) \sin^2 \theta \tag{27}$$

The flow pattern is given in Fig. 32.

36. Fluid Motion about a Stationary Sphere Due to Movement of a Concentric Spherical Shell. The preceding case may be converted into one in which the sphere is stationary and the shell has a velocity U in the negative x-direction.[1] This is accomplished by adding to the velocity potential and stream function a uniform flow given by

$$\phi = Ur \cos \theta, \qquad \psi = \frac{Ur^2}{2} \sin^2 \theta$$

respectively. The equations become

$$\phi = \frac{a^3 U}{b^3 - a^3} \left(r + \frac{b^3}{2r^2} \right) \cos \theta + Ur \cos \theta$$

$$\psi = \frac{a^3 U}{2(b^3 - a^3)} \left(r^2 - \frac{b^3}{r} \right) \sin^2 \theta + \frac{Ur^2}{2} \sin^2 \theta$$

when $\psi = 0$, $\theta = 0$, and $r = a$ and

when $r = b$, $\dfrac{\partial \phi}{\partial r} = U \cos \theta$. The velocity is necessarily tangent to the sphere and is given by

$$q \Big]_{r=a} = \left(1 + \frac{2a^3 + b^3}{2(b^3 - a^3)} \right) U \sin \theta$$

when b approaches a the velocity approaches infinity. For very large values of b this reduces to the case in Sec. 34. The flow pattern is given in Fig. 33.

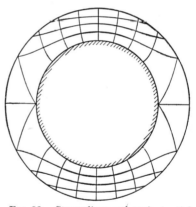

Fig. 33.—Streamlines and equipotential lines for motion of a fluid about a sphere due to movement of a concentric spherical shell.

37. Flow Due to a Doublet Near a Sphere. The flow net resulting from two unequal doublets oppositely directed along the same axis contains a closed spherical stream surface about one of the doublets. To prove this the doublets are located at O and R and a coordinate system

---

[1] G. G. Stokes, "Mathematical and Physical Papers," Vol. 1, pp. 230–235, Cambridge University Press, London, 1880.

selected as shown in Fig. 34. From Eq. (18), a doublet with axis in the positive x-direction is given by

$$\psi = -\frac{\mu\hat{\omega}^2}{r^3}$$

Hence, the stream function for oppositely directed doublets is

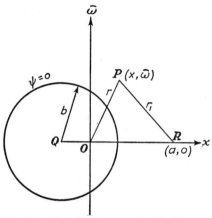

$$\psi = \frac{\mu\hat{\omega}^2}{r^3} - \frac{\mu_1\hat{\omega}^2}{r_1{}^3} \qquad (28)$$

where r, r_1 denote the distances of doublets from any point P. The streamline $\psi = 0$ is given by

$$\frac{r}{r_1} = \left(\frac{\mu}{\mu_1}\right)^{\frac{1}{3}} = c \qquad (29)$$

Expressing r and r_1 in terms of $\hat{\omega}$ and x, the equation may take the form

$$\left(x + \frac{c^2 a}{1 - c^2}\right)^2 + \hat{\omega}^2 = \frac{a^2 c^2}{(1 - c^2)^2}$$

When $\mu_1 > \mu$, this is a sphere of

Fig. 34.—Notation for opposite doublets at O and R.

radius $b = ac/(1 - c^2)$, or

$$b = \frac{\overline{OR}\mu^{\frac{1}{3}}\mu_1{}^{\frac{1}{3}}}{\mu_1{}^{\frac{2}{3}} - \mu^{\frac{2}{3}}}$$

and with center Q at

$$x = -\frac{c^2 a}{1 - c^2} = -\overline{QO} = -\overline{OR}\frac{\mu^{\frac{2}{3}}}{\mu_1{}^{\frac{2}{3}} - \mu^{\frac{2}{3}}}, \qquad \hat{\omega} = 0$$

O and R are inverse points in the sphere; *i.e.*,

$$\overline{QO} \times \overline{QR} = b^2$$

as can easily be shown by substitution. When $\mu > \mu_1$, the sphere encloses R.

38. Flow Due to a Source Near a Sphere. The combination of a source of strength m at P (Fig. 35), another source of strength $(m\,\overline{OQ})/a$ at Q, and a uniform line sink, of strength m/a per unit length, from O to Q produces a closed spherical stream surface of radius a with center at O, provided P and Q are inverse points with respect to the sphere.[1]

[1] W. M. Hicks, On the Motion of Two Spheres in a Fluid, *Phil. Trans. Roy. Soc. London*, Vol. 171, pp. 455–492, 1880.

The stream function for the line sink (Sec. 31) is

$$\psi = -\frac{m}{4\pi a}(\overline{RO} - \overline{RQ})$$

The stream function for the source at P, from Eq. (2), is

$$\psi = \frac{m}{4\pi}\cos RPx$$

and for the source at Q,

$$\psi = \frac{m}{4\pi a}\overline{OQ}\cos RQP$$

Adding the three stream functions,

$$\psi = \frac{m}{4\pi a}(\overline{RQ} - \overline{RO}) + \frac{m}{4\pi}\cos RPx + \frac{m\overline{OQ}}{4\pi a}\cos RQP$$

Restricting R to the sphere of radius a,

$$\overline{RQ} = \overline{RO}\cos ORQ + \overline{OQ}\cos OQR, \qquad \cos RPx = -\cos RPO$$
$$\cos RQP = -\cos RQO, \qquad \text{angle } RPO = \text{angle } ORQ$$

The latter equality follows from the triangles ROQ and ROP, which are similar because they have a common side and angle, and because

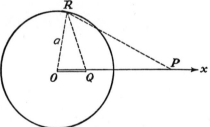

$$\frac{\overline{OQ}}{a} = \frac{a}{\overline{OP}}$$

since P is the inverse of Q in the sphere. Substituting into the stream function it reduces to

$$\psi = -\frac{m}{4\pi}$$

Fig. 35.—Combinations of two point sources at P and Q and a line sink.

over the sphere; hence, the sphere is a stream surface that encloses the source at Q and the line sink. Therefore, the resulting pattern from the stream function is due to a source near a sphere.

39. Surface Zonal Harmonics. In several examples of fluid flow encountered in this chapter the type or form of the velocity potential has been assumed and then shown to satisfy the boundary conditions for a given situation. In dealing with sources and doublets the term $1/R$

(Fig. 36) occurs and may be expressed in terms of r, θ, and $c = \overline{OA}$. Using the cosine law,

$$R^2 = r^2 + c^2 - 2cr \cos \theta = r^2 \left[1 - 2 \left(\frac{c}{r} \right) \cos \theta + \left(\frac{c}{r} \right)^2 \right]$$

$$= c^2 \left[1 - 2 \left(\frac{r}{c} \right) \cos \theta + \left(\frac{r}{c} \right)^2 \right]$$

Writing either of the latter two terms in the form

$$1 - 2\lambda \cos \theta + \lambda^2$$

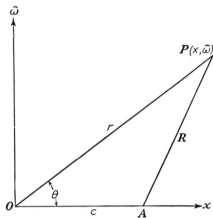

FIG. 36.—Relationships among r, c, R, and θ in surface zonal harmonics.

the following expansion can be made:

$$(1 - 2\lambda \cos \theta + \lambda^2)^{-\frac{1}{2}}$$
$$= P_0 (\cos \theta) + \lambda P_1 (\cos \theta) + \lambda^2 P_2 (\cos \theta) + \lambda^3 P_3 (\cos \theta) + \cdots$$

where P_0, P_1, P_2, . . . are independent of λ. Performing the expansion and solving for P_0, P_1, P_2, P_3, . . .

$$P_0 = 1$$
$$P_1 = \cos \theta$$
$$P_2 = \tfrac{1}{2}(3 \cos^2 \theta - 1)$$
$$P_3 = \tfrac{1}{2}(5 \cos^3 \theta - 3 \cos \theta)$$

The functions P_0, P_1, P_2, P_3 are known as *surface zonal harmonics*. Expressing $1/R$ in terms of zonal harmonics,

$$\frac{1}{R} = \frac{1}{c} + \frac{r}{c^2} P_1(\cos \theta) + \frac{r^2}{c^3} P_2(\cos \theta) + \frac{r^3}{c^4} P_3(\cos \theta) + \cdots$$

which is valid for $r < c$. When $r > c$, then

$$\frac{1}{R} = \frac{1}{r} + \frac{c}{r^2} P_1(\cos \theta) + \frac{c^2}{r^3} P_2(\cos \theta) + \frac{c^3}{r^4} P_3(\cos \theta) + \cdots$$

It may be shown by substitution in $\nabla^2 \phi = 0$ that each term of either expression for $1/R$ satisfies the Laplace equation. This may also be reasoned from the fact that the expansion $1/R$ in series must satisfy $\nabla^2 \phi = 0$ as a whole; then as no two terms are of the same degree and thus cannot cancel, each term in itself must be a solution. Therefore, the expressions in terms of surface zonal harmonics provide an infinite number of solutions of the Laplace equation in three dimensions.

Examples of uses in the preceding sections are as follows:

Source at origin:

$$\phi = C\frac{P_0}{r}$$

Uniform flow in x direction:

$$\phi = CrP_1$$

Half body:

$$\phi = C_1 r P_1 + C_2 \frac{P_0}{r}$$

Translation of sphere through infinite fluid, or doublet:

$$\phi = C\frac{P_1}{r^2}$$

Steady flow of infinite fluid around a sphere:

$$\phi = C_1 r P_1 + C_2 \frac{P_1}{r^2}$$

40. Fluid Motion Due to Two Spheres Moving along Their Line of Centers.[1] Consider two spheres in an infinite fluid (Fig. 37) moving in their line of centers. The velocity potential for the fluid surrounding these two spheres is sought. This flow case has axial symmetry with respect to the line of centers; hence, a point P may be specified by r and θ or by r' and θ'. Letting ϕ_0 be the desired velocity potential, the boundary conditions at the surfaces of the spheres are

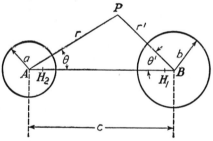

Fig. 37.—Notation for two spheres moving along their line of centers.

$$\frac{\partial \phi_0}{\partial r}\bigg]_{r=a} = -U\cos\theta, \qquad \frac{\partial \phi_0}{\partial r'}\bigg]_{r'=b} = -U'\cos\theta'$$

where U is the velocity of sphere A toward B, and U' the velocity of sphere B toward A.

Writing the velocity potential in the form

$$\phi_0 = U\phi + U'\phi' \tag{30}$$

[1] For references on the motion of two spheres see Lamb, "Hydrodynamics," 6th ed., p. 134, Cambridge University Press, London, 1932; see also reference under Sec. 38.

the velocity potential ϕ is for fluid motion for sphere B stationary as sphere A moves with unit velocity toward B. ϕ' may be given a similar interpretation. The boundary conditions now become

$$\left.\frac{\partial\phi}{\partial r}\right]_{r=a} = -\cos\theta \tag{a}$$

$$\left.\frac{\partial\phi'}{\partial r}\right]_{r=a} = 0 \tag{b}$$

$$\left.\frac{\partial\phi}{\partial r'}\right]_{r'=b} = 0 \tag{c}$$

$$\left.\frac{\partial\phi'}{\partial r'}\right]_{r'=b} = -\cos\theta' \tag{d}$$

The potential functions ϕ, ϕ' are independently built up by taking successive images of doublets in the two spheres at inverse points, determining location and strength by the methods of Sec. 37.

To obtain ϕ, first consider B absent. The velocity potential for a sphere at A of radius a moving with unit velocity toward B is

$$\phi_1 = \frac{1}{2}\frac{a^3}{r^2}\cos\theta$$

from Sec. 33. This satisfies condition (a), but condition (c) is not satisfied. Introducing a doublet at point H_1 in B, where

$$\overline{H_1B} = \frac{b^2}{c}$$

satisfies the boundary condition over sphere B, provided the strength of doublet at H_1 is given [see Eq. (29), Sec. 37] by

$$\frac{b - \overline{H_1B}}{c - b} = \left(\frac{\mu_1}{a^3/2}\right)^{\frac{1}{3}}$$

where μ_1 is the strength of doublet at H_1 and $\mu_0 = a^3/2$ is the strength of doublet at A. Eliminating $\overline{H_1B}$ in the two expressions and simplifying,

$$\mu_1 = \frac{a^3b^3}{2c^3}$$

The doublet at H_1 has its axis directed toward A. Letting r_1 be the distance from P to H_1 and $\cos\theta_1$ be $\cos PH_1A$,

$$\phi_2 = \frac{1}{2}\frac{a^3}{r^2}\cos\theta - \frac{1}{2}\frac{a^3b^3}{c^3r_1^2}\cos\theta_1$$

is an approximate solution satisfying condition (c) but not condition (a), due to effect of the latter term.   Another doublet in sphere A at H_2, which is the inverse of H_1 in sphere A and of strength

$$\mu_2 = \frac{a^6 b^3}{2c^3 \overline{AH_1}^3}$$

satisfies condition (a) again but unbalances condition (c).   Continuing this procedure the potential function may be obtained to any desired degree of accuracy.   The first four terms of ϕ are

$$\phi = \frac{a^3}{2r^2} \cos \theta + \frac{a^3 b^3}{2c^3 r_1^2} \cos \theta_1 + \frac{a^6 b^3}{2c^3 f_1^3} \frac{\cos \theta_2}{r_2^2} + \frac{1}{2} \frac{a^6 b^6 \cos \theta_3}{c^3 f_1^3 (c - f_2)^3 r_3^2}$$

where $f_1, f_2, \ldots$ represent the distances of successive images from A, i.e., $f_n = AH_n$, $n = 1, 2, 3, \ldots$; $\theta_1, \theta_2, \theta_3, \ldots$ represent the angle $PH_1, PH_2, PH_3, \ldots$ make with line of centers; $r_1, r_2, r_3, \ldots$ represent the distances $\overline{PH_1}, \overline{PH_2}, \overline{PH_3}, \ldots$.

The potential function ϕ' is developed in an analogous manner, starting with a doublet strength $b^3/2$ at B.

The equation is made up from a series of doublets, any of which satisfy $\nabla^2 \phi = 0$; therefore their sum is also a solution.   The strength of the doublets decreases rapidly with each term, particularly when c is large compared with a and b.   The velocity becomes zero at infinity; hence, all necessary conditions are satisfied.   Computation is awkward with the exact solution.   An approximate solution more amenable to calculation is as follows:

Assuming B absent,

$$\phi = \frac{1}{2} \frac{a^3}{r^2} \cos \theta$$

satisfies condition (a) as before.   To obtain an approximate form for ϕ in the vicinity of B, as $r \cos \theta = c - r' \cos \theta'$,

$$\phi = \frac{a^3}{2r^3} (c - r' \cos \theta')$$

$$= \frac{a^3}{2} \frac{c - r' \cos \theta'}{(c^2 - 2r'c \cos \theta' + r'^2)^{3/2}}$$

Expanding the term for $1/r^3$ in series

$$\phi = \frac{a^3}{2} (c - r' \cos \theta') \left(\frac{1}{c^3} + \frac{3r'}{c^4} \cos \theta' - \frac{3}{2} \frac{r'^2}{c^5} + \cdots \right)$$

$$\cong \frac{a^3}{2c^2} \left(1 + \frac{2r' \cos \theta'}{c} \right)$$

near B.   The normal velocity over B is

$$-\frac{\partial \phi}{\partial r'}\bigg]_{r'=b} = -\frac{a^3}{c^3}\cos \theta'$$

This velocity is canceled by the addition of

$$\frac{a^3 b^3}{2c^3}\frac{\cos \theta'}{r'^2}$$

which has a value near A, by same procedure as above, of

$$\frac{a^3 b^3}{2c^5}\left(1 + \frac{2r \cos \theta}{c}\right)$$

The normal velocity over A due to this term is

$$-\frac{\partial \phi}{\partial r}\bigg]_{r=a} = -\frac{a^3 b^3}{c^6}\cos \theta$$

which may be canceled by addition of the velocity potential

$$\frac{a^6 b^3}{2c^6}\frac{\cos \theta}{r^2}$$

The velocity potential near A to this degree of approximation is

$$\phi = \frac{a^3}{2r^2}\cos \theta + \frac{a^3 b^3}{2c^5}\left(1 + \frac{2r \cos \theta}{c}\right) + \frac{a^6 b^3}{2c^6}\frac{\cos \theta}{r^2}$$

and the velocity potential near B is

$$\phi = \frac{a^3}{2c^2}\left(1 + \frac{2r' \cos \theta'}{c}\right) + \frac{a^3 b^3}{2c^3}\frac{\cos \theta'}{r'^2}$$

The velocity potential for ϕ' is found similarly.

In order to evaluate the kinetic energy of the fluid, using Eq. (11), Sec. 14, the values of ϕ, ϕ' and $\dfrac{\partial \phi}{\partial n}$, $\dfrac{\partial \phi'}{\partial n}$ are obtained; thus

$$\phi = \tfrac{1}{2}a\left(1 + \frac{3a^3 b^3}{c^6}\right)\cos \theta + \text{constant} \quad \text{(on sphere } A\text{)}$$

$$\phi' = \tfrac{1}{2}b\left(1 + \frac{3a^3 b^3}{c^6}\right)\cos \theta' + \text{constant} \quad \text{(on sphere } B\text{)}$$

$$\phi = \frac{a^3}{2c^2}\left(1 + \frac{3b}{c}\cos \theta'\right) \quad\quad \text{(on sphere } B\text{)}$$

$$\phi' = \frac{b^3}{2c^2}\left(1 + \frac{3a}{c}\cos \theta\right) \quad\quad \text{(on sphere } A\text{)}$$

and from the boundary conditions

$$\frac{\partial \phi}{\partial n} = -\cos \theta, \qquad \frac{\partial \phi'}{\partial n} = -\cos \theta'$$

over spheres A and B, respectively.   Then

$$\int \phi_0 \frac{\partial \phi_0}{\partial n}\, dS = U^2 \int \phi \frac{\partial \phi}{\partial n}\, dS_A + UU' \int \phi' \frac{\partial \phi}{\partial n}\, dS_A$$

$$+ UU' \int \phi \frac{\partial \phi'}{\partial n}\, dS_B + U'^2 \int \phi' \frac{\partial \phi'}{\partial n}\, dS_B$$

where dS_A, dS_B represent area elements on spheres A and B.   Integrating the first term on the right, using the area element $dS_A = 2\pi a^2 \sin \theta\, d\theta$,

$$U^2 \int \phi \frac{\partial \phi}{\partial n}\, dS_A = -U^2 \pi a^3 \left(1 + \frac{3a^3 b^3}{c^6}\right) \int_0^\pi \cos^2 \theta \sin \theta\, d\theta$$

$$= -\tfrac{2}{3}\pi a^3 U^2 \left(1 + \frac{3a^3 b^3}{c^6}\right)$$

The constant term in ϕ cannot contribute to the integral, since

$$\int \frac{\partial \phi}{\partial n}\, dS = 0$$

by the continuity equation.   Integrating the second term,

$$UU' \int \phi' \frac{\partial \phi}{\partial n}\, dS_A = -\frac{2\pi a^3 b^3}{c^3}\, UU'$$

The third integral is the same as the second one, and the last gives

$$-\tfrac{2}{3}\pi b^3 U'^2 \left(1 + \frac{3a^3 b^3}{c^6}\right)$$

The kinetic energy of the fluid is

$$T = 2\pi\rho \frac{a^3 b^3}{c^3}\, UU' + \frac{\pi\rho}{3}(a^3 U^2 + b^3 U'^2)\left(1 + \frac{3a^3 b^3}{c^6}\right)$$

or

$$T = \left(\frac{M_A}{4} U^2 + \frac{M_B}{4} U'^2\right)\left(1 + \frac{3a^3 b^3}{c^6}\right) + 2\pi\rho \frac{a^3 b^3}{c^3}\, UU'$$

where M_A, M_B are the masses of fluid displaced by the respective spheres.

41. Fluid Motion Due to Sphere Moving Perpendicularly to a Fixed Plane Boundary.   When $a = b$ and $U = U'$ in the preceding section, the plane that bisects line AB is a plane of symmetry across which there is no flow.   It can, therefore, be replaced by an infinite rigid boundary, resulting in a flow case where a sphere moves with velocity U toward a wall. The velocity potential is given approximately by

$$\phi = \frac{U}{2}\left(\frac{a^3}{r^2} + \frac{a^3 r}{4h^3} + \frac{a^6 r}{32h^6} + \frac{a^6}{8h^3 r^2} + \frac{a^9}{64h^6 r^2}\right)\cos\theta + \text{constant}$$

The stream function is

$$\psi = \frac{U}{4}\left(-\frac{2a^3}{r} + \frac{a^3 r^2}{4h^3} + \frac{a^6 r^2}{32h^6} - \frac{a^6}{4h^3 r} - \frac{a^9}{32h^6 r}\right)\sin^2\theta + \text{constant}$$

where $h = c/2$.

The kinetic energy of the fluid is

$$T = \pi\rho a^3 U^2\left(\frac{1}{3} + \frac{a^3}{8h^3} + \frac{a^6}{64h^6}\right)$$

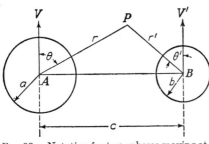

Letting M represent the mass of the sphere and M' the mass of displaced fluid, the total kinetic energy of fluid and sphere is

$$\tfrac{1}{4}\left(2M + M' + \tfrac{3}{8}M'\frac{a^3}{h^3}\right.$$

$$\left. + \tfrac{3}{64}M'\frac{a^6}{h^6}\right)U^2$$

Fig. 38.—Notation for two spheres moving at right angles to their line of centers.

As the sphere approaches the wall, under the action of no external forces, the kinetic energy of the system must remain constant; hence U decreases as h decreases. Similarly, if the sphere moves away from the wall, U increases. The sphere is repelled by the wall in both cases. This phenomenon decreases the probability of a head-on collision between immersed or floating objects.

42. Fluid Motion Due to Two Spheres Moving at Right Angles to Their Line of Centers. An approximate solution of this flow case will be derived in a manner similar to that in Sec. 40. Referring to Fig. 38 for notation, as before let

$$\phi_0 = V\phi + V'\phi'$$

where ϕ is the velocity potential for unit velocity of sphere A while sphere B is motionless and ϕ' is the velocity potential for unit velocity of sphere B while sphere A is motionless. V and V' must be parallel. The boundary conditions at the surfaces of the spheres are

$$\frac{\partial \phi}{\partial r}\bigg]_{r=a} = -\cos \theta \qquad (a)$$

$$\frac{\partial \phi'}{\partial r}\bigg]_{r=a} = 0 \qquad (b)$$

$$\frac{\partial \phi}{\partial r'}\bigg]_{r'=b} = 0 \qquad (c)$$

$$\frac{\partial \phi'}{\partial r'}\bigg]_{r'=b} = -\cos \theta' \qquad (d)$$

The velocity potential for A, if B were absent, is

$$\frac{a^3}{2r^2}\cos \theta$$

which satisfies condition (a).   From Fig. 38

$$r \cos \theta = r' \cos \theta'$$

therefore, the velocity potential in the neighborhood of B may be written

$$\frac{a^3}{2c^3} r' \cos \theta'$$

assuming that $r = c$ does not materially change the velocity potential.
The normal velocity over the surface of B is

$$-\frac{a^3}{2c^3}\cos \theta'$$

which may be canceled by addition of

$$\frac{a^3 b^3}{4c^3}\frac{\cos \theta'}{r'^2}$$

to the velocity potential near B.   In the vicinity of A this may be
approximated by

$$\frac{a^3 b^3}{4c^6} r \cos \theta$$

by assuming $r' = c$ over the surface of A.   To cancel the velocity over A
due to this term it is necessary to add

$$\frac{a^6 b^3}{8c^6}\frac{\cos \theta}{r^2}$$

Neglecting terms with powers of c higher than c^{-3}, the velocity potential becomes

$$\phi = \frac{a^3}{2}\frac{\cos \theta}{r^2} + \frac{a^3 b^3}{4c^3}\frac{\cos \theta'}{r'^2}$$

Hence, over the surface of A

$$\phi = \frac{a}{2} \cos \theta$$

while over the surface of B

$$\phi = \frac{3a^3 b}{4c^3} \cos \theta'$$

In an analogous manner the velocity potential ϕ' for unit velocity of B and zero velocity of A is

$$\phi' = \frac{b^3}{2} \frac{\cos \theta'}{r'^2} + \frac{a^3 b^3}{4c^3} \frac{\cos \theta}{r^2}$$

which becomes over the surface of B

$$\phi' = \frac{b}{2} \cos \theta'$$

neglecting terms containing c^{-6} and higher.   Over surface A

$$\phi' = \frac{3}{4} \frac{ab^3}{c^3} \cos \theta$$

Collecting results, near A

$$\phi_0 = \frac{a^3 \cos \theta}{2r^2} \left(V + \frac{3}{2} V' \frac{b^3}{c^3} \right)$$

while near B

$$\phi_0 = \frac{b^3 \cos \theta'}{2r'^2} \left(V' + \frac{3}{2} V \frac{a^3}{c^3} \right)$$

All equations are accurate only when c is great compared with a and b and only when the spheres are moving at right angles to their line of centers AB.

Using the same method of evaluating kinetic energy of fluid as in Sec. 40,

$$T = \frac{\pi \rho a^3}{3} V^2 + \pi \rho \frac{a^3 b^3}{c^3} VV' + \frac{\pi \rho b^3}{3} V'^2$$

43. Fluid Motion Due to Sphere Moving Parallel to a Fixed Plane Boundary.   This is a special case of the preceding section, where $a = b$, $V = V'$, and, as before, where $c \gg a$.   The plane bisecting AB becomes a plane of symmetry across which there is no flow.   It may, therefore, be replaced by a rigid plane boundary.   Let $c = 2h$; then

$$\phi_0 = \frac{a^3 \cos \theta}{2r^2} V \left(1 + \frac{3}{16} \frac{a^3}{h^3} \right)$$

near the sphere.   The kinetic energy of the fluid becomes

$$T - \frac{\pi \rho a^3}{3} \left(1 + \frac{3}{16} \frac{a^3}{h^3}\right) V^2$$

where h is the distance from center of sphere to wall.

44. Motion of Fluid Bounded Externally by an Ellipsoid.   When dealing with fluid flow cases in which the fluid has external boundaries, two conditions only must be satisfied, *viz.*,

$$\nabla^2 \phi = 0 \quad \text{and} \quad \frac{DF}{Dt} = 0$$

The equation of the ellipsoid of semiaxes a,b,c, moving uniformly with velocity components U,V,W in the positive xyz-directions, respectively, is

$$F = \frac{(x - Ut)^2}{a^2} + \frac{(y - Vt)^2}{b^2} + \frac{(z - Wt)^2}{c^2} - 1 = 0$$

For $t = 0$, the boundary condition equation $\frac{DF}{Dt} = 0$ becomes

$$\frac{DF}{Dt} = \frac{ux}{a^2} + \frac{vy}{b^2} + \frac{wz}{c^2} - \frac{Ux}{a^2} - \frac{Vy}{b^2} - \frac{Wz}{c^2} = 0$$

In terms of the velocity potential this becomes

$$\frac{x}{a^2} \frac{\partial \phi}{\partial x} + \frac{y}{b^2} \frac{\partial \phi}{\partial y} + \frac{z}{c^2} \frac{\partial \phi}{\partial z} + \frac{x}{a^2} U + \frac{y}{b^2} V + \frac{z}{c^2} W = 0$$

It may be seen by inspection that this equation and the Laplace equation are both satisfied by

$$\phi = -Ux - Vy - Wz$$

Hence, by Sec. 15, the only possible irrotational motion of a uniformly translating ellipsoid occurs when the fluid is moving as if it were a solid having the same velocity as the ellipsoidal shell.

The case of rotation of the general ellipsoid about the x-axis is investigated where the fluid is contained within the ellipsoidal shell.   To develop the boundary condition that must be satisfied for this case the fluid velocity normal to the surface of the ellipsoid is found and equated to the velocity of the surface normal to itself.   First, to find the direction cosines of the normal to the surface [Eq. (12), Sec. 8],

$$F = \frac{x^2}{a^2} + \frac{y^2}{b^2} + \frac{z^2}{c^2} - 1 = 0$$

Differentiating,

$$\frac{\partial F}{\partial x} = \frac{2x}{a^2}, \qquad \frac{\partial F}{\partial y} = \frac{2y}{b^2}, \qquad \frac{\partial F}{\partial z} = \frac{2z}{c^2}$$

and

$$l, m, n = \frac{-\dfrac{x}{a^2}, -\dfrac{y}{b^2}, -\dfrac{z}{c^2}}{\sqrt{\dfrac{x^2}{a^4} + \dfrac{y^2}{b^4} + \dfrac{z^2}{c^4}}}$$

for the cosines of the normal drawn into the fluid.   The velocity of the fluid in the direction of this normal is

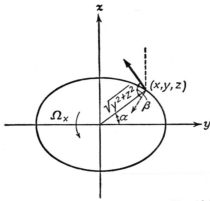

$$lu + mv + nw$$
$$= \frac{\dfrac{x}{a^2}\dfrac{\partial \phi}{\partial x} + \dfrac{y}{b^2}\dfrac{\partial \phi}{\partial y} + \dfrac{z}{c^2}\dfrac{\partial \phi}{\partial z}}{\sqrt{\dfrac{x^2}{a^4} + \dfrac{y^2}{b^4} + \dfrac{z^2}{c^4}}} \qquad (31)$$

The velocity of any point (x,y,z) of the boundary, due to the rotation Ω_x about the x-axis, is, from Fig. 39, $\Omega_x \sqrt{y^2 + z^2}$ in magnitude, with direction cosines l_1, m_1, n_1, where

$$l_1 = 0, \qquad m_1 = \frac{-z}{\sqrt{y^2 + z^2}},$$

$$n_1 = \frac{y}{\sqrt{y^2 + z^2}}$$

FIG. 39.—Velocity of a point on an ellipsoidal shell rotating about a principal axis.

Let β represent the angle between the velocity vector and normal, then the velocity of the surface normal to itself is

$$\Omega_x \sqrt{y^2 + z^2} \cos \beta = \Omega_x \sqrt{y^2 + z^2}\,(ll_1 + mm_1 + nn_1)$$

$$= \frac{\Omega_x yz}{\sqrt{\dfrac{x^2}{a^4} + \dfrac{y^2}{b^4} + \dfrac{z^2}{c^4}}}\left(\frac{1}{b^2} - \frac{1}{c^2}\right) \qquad (32)$$

From Eqs. (31) and (32) the boundary condition is obtained:

$$\frac{x}{a^2}\frac{\partial \phi}{\partial x} + \frac{y}{b^2}\frac{\partial \phi}{\partial y} + \frac{z}{c^2}\frac{\partial \phi}{\partial z} = \Omega_x yz\left(\frac{1}{b^2} - \frac{1}{c^2}\right)$$

Assume

$$\phi = Cyz$$

which satisfies $\nabla^2 \phi = 0$.   It is found to be the solution, provided

$$C = \frac{c^2 - b^2}{c^2 + b^2}\,\Omega_x$$

Hence,

$$\phi = \frac{c^2 - b^2}{c^2 + b^2} \, \Omega_x yz$$

is the potential function for fluid motion caused by the rotation of an ellipsoidal shell about the x-axis, when the shell is completely filled with fluid.

Due to lack of symmetry, the Stokes' stream function is not defined for this case.

Exercises

1. Select the proper strength of source and uniform flow to produce a dynamic pressure of 100 lb per ft² at the stagnation point on a half body whose asymptotic cylinder is $\bar{\omega} = 2$ ft. The fluid has a mass density 2.00 slugs per ft³. Draw the half body and the lines of constant pressure: $p = 75$ lb per ft², $p = 50$ lb per ft², $p = 25$ lb per ft². *Ans.* $U = 10$ ft per sec; $m = 40\pi$ ft³ per sec.

2. Find the point of minimum pressure in the fluid surrounding the half body of Exercise 1. What is this minimum pressure?

3. (*a*) Draw the traces of three stream surfaces about the half body of Exercise 1 for a $2\pi \, \Delta\psi$ of 25 ft³ per sec.

(*b*) Draw the three equipotential lines $\phi_0 = 25$, $\phi_1 = 5$, $\phi_2 = -15$.

4. By integration over the surface of a half body show that its drag is zero.

5. Using Eq. (20), Sec. 17, derive the stream function, Eq. (11), from Eq. (10) for a source and sink of equal strength.

6. Why cannot point D of Fig. 22 be found from Eq. (15)?

7. (*a*) Find the equations for flow around a Rankine body 3 ft long and 2 ft in breadth.

(*b*) What is the pressure at $x = 0$, $\bar{\omega} = 1$ for $U = 3.0$ ft per sec, liquid of density $\rho = 1.94$ slugs per ft³, taking dynamic pressure zero at infinity?

8. Obtain Eq. (18) by applying the limiting process to Eq. (11).

9. Draw a curve of constant pressure intensity around a doublet. Take

$$\mu = 100 \text{ ft}^4 \text{ per sec,}$$

$P = -50$ lb per ft², $\rho = 2$ slugs per ft³ (pressure zero at infinity).

10. Find the length and breadth of the body formed by the closing stream surface when there is a line sink of strength $m = 20$ ft³ per sec per ft extending from $(-1,0)$ to $(0,0)$, a line source of the same strength from $(0,0)$ to $(1,0)$, and a uniform flow in the negative x-direction of 5 ft per sec.

11. If the dynamical pressure at infinity is zero in Exercise 10, find the pressure at $(0,2)$.

12. Derive the Laplace equation for the cylindrical coordinates $(x, \bar{\omega})$ when there is axial symmetry.

13. If a sphere 1.0 ft in diameter and having a unit weight of 450 lb per ft³ is released in water at rest, find its velocity after 2 sec. Assume irrotational flow of an ideal fluid.

14. Show by taking the surface integral of the pressure force in the x-direction that a sphere at rest in a uniform, ideal fluid stream suffers no drag force.

15. Where should the openings be placed in a spherical Pitot tube to measure static pressure in the fluid?

16. Find the virtual mass for the sphere moving inside a fixed concentric spherical shell.   *Ans.* $\dfrac{2}{3}\pi\dfrac{b^3+2a^3}{b^3-a^3}\rho a^3.$

17. Show by substitution that $r^3 P_3$ (cos θ) is a solution of $\nabla^2\phi = 0$.

18. A 100-lb sphere, 1.0 ft in diameter, is submerged in water and released from rest when its center is 6.0 ft from a horizontal floor. Assuming ideal fluid flow, what velocity does the sphere have when it has fallen 4.0 ft?

CHAPTER V

APPLICATION OF COMPLEX VARIABLES
TO TWO-DIMENSIONAL FLUID FLOW

An infinite number of solutions to the two-dimensional Laplace equation are easily obtainable from functions of a complex variable. In fact, due to the application of complex variable theory, the study of ideal fluid flow has been greatly expanded. Elements of complex variable theory are introduced in the first part of the chapter and conformal mapping in the remaining part.

COMPLEX VARIABLES

45. Complex Numbers. When solving quadratic equations the discriminant may be negative, giving a solution of the form

$$x + iy$$

where x,y are real numbers and $i = \sqrt{-1}$; i.e., $i^2 = -1$. The quantity $x + iy$ is called a *complex number*. It is made up of two parts: the real part x and the pure imaginary part y.

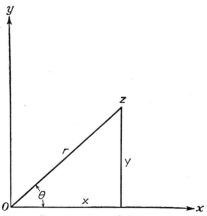

The Argand diagram (Fig. 40) provides a convenient geometric representation of complex numbers. In the Argand diagram the real part of the complex number is plotted as abscissa against the pure imaginary part as ordinate. For example, the complex number $x + iy$ is represented by the line Oz extending from the origin to the point (x,y) on the graph. The

FIG. 40.—Argand diagram.

concept of the complex number is more easily grasped by considering it as a directed line segment rather than a point on a plane. Thus, in the figure, the complex number has a length r and makes an angle θ with the x-axis. The length of the line representing it is known as the *modulus*, or absolute value of the complex number, which may be designated by r or by $|z|$, where

$$z = x + iy$$

85

Therefore,

$$|z| = \sqrt{x^2 + y^2}$$

A complex number may conveniently be specified by using polar coordinates. Since $x = r \cos \theta$, $y = r \sin \theta$,

$$z = r \cos \theta + ir \sin \theta = r(\cos \theta + i \sin \theta)$$

where r is the modulus and θ is the angle the radius vector makes with the x-axis measured positive in the counterclockwise direction and is known as the *argument* of the complex number. A complex number is completely specified when its modulus and argument are given. Two complex numbers are equal if their moduli are the same and their arguments are the same. Similarly, from the cartesian notation, two complex numbers are equal if their real parts are equal and if their pure imaginary parts are equal.

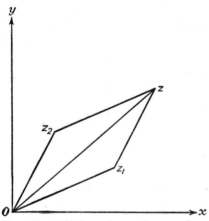

Complex numbers are added according to the parallelogram law of adding vectors, as indicated in Fig. 41. There,

FIG. 41.—Addition of complex numbers.

$$z = z_1 + z_2 = x_1 + x_2 + i(y_1 + y_2)$$

This is equivalent to adding their real parts and to adding their pure imaginary parts. The difference of two complex numbers is obtained by taking the difference of their real parts and the difference of their pure imaginary parts; thus

$$z - z_1 = x - x_1 + i(y - y_1) = z_2$$

from Fig. 41.

The factor i that enters into complex numbers may be considered as an operator such that multiplication by it rotates a complex number through $+90$ deg. For example, when $z = 0 + 3i$ is multiplied by i, the result is $iz = 0 - 3$, indicating rotation from the Oy-axis to the negative x-axis. In more general form,

$$z = a + ib, \qquad iz = -b + ia$$

the slope of z is b/a and the slope of iz, $-a/b$, showing they are at right angles. The rotation may be shown to be in the counterclockwise direction by plotting or by using the exponential form derived in the next section. Application of i twice is equivalent to rotation through 180 deg, since $i^2 = -1$.

46. DeMoivre's Theorem. A complex number may also be expressed in an exponential form, which is in every way equivalent to the cartesian and polar forms but more suitable for certain operations. The relationship is known as *DeMoivre's theorem* and is derived from a consideration of infinite series expansions of e^x, sin x, and cos x. The Maclaurin series expansion[1] for e^x is

$$e^x = 1 + x + \frac{x^2}{2!} + \frac{x^3}{3!} + \frac{x^4}{4!} + \cdots$$

and the series expansions for sin x and cos x are

$$\sin x = x - \frac{x^3}{3!} + \frac{x^5}{5!} - \frac{x^7}{7!} + \cdots$$

$$\cos x = 1 - \frac{x^2}{2!} + \frac{x^4}{4!} - \frac{x^6}{6!} + \cdots$$

All the above series are absolutely convergent for all values of x. Replacing x by $i\theta$ in the expression e^x and remembering that $i^2 = -1$, $i^3 = -i$, $i^4 = 1$, etc.

$$e^{i\theta} = 1 + i\theta - \frac{\theta^2}{2!} - i\frac{\theta^3}{3!} + \frac{\theta^4}{4!} + i\frac{\theta^5}{5!} - \frac{\theta^6}{6!} - \cdots$$

which will be taken as the definition of $e^{i\theta}$.

Since the series is absolutely convergent, it can be rearranged; thus

$$e^{i\theta} = 1 - \frac{\theta^2}{2!} + \frac{\theta^4}{4!} - \frac{\theta^6}{6!} + \cdots + i\left(\theta - \frac{\theta^3}{3!} + \frac{\theta^5}{5!} - \frac{\theta^7}{7!} + \cdots\right)$$

which is seen to be

$$e^{i\theta} = \cos\theta + i\sin\theta \tag{1}$$

by replacing x by θ in the sine and cosine series. Equation (1) is DeMoivre's theorem.

The three forms available for expression of a complex number are

$$z = x + iy = r(\cos\theta + i\sin\theta) = re^{i\theta} \tag{2}$$

where the most convenient form may be selected. The terms may be treated as ordinary algebraic quantities subject only to $i^2 = -1$. It follows that

$$z^n = r^n e^{in\theta} = r^n(\cos n\theta + i\sin n\theta)$$

[1] I. S. and E. S. Sokolnikoff, "Higher Mathematics for Engineers and Physicists," 2d ed., p. 37, McGraw-Hill Book Company, Inc., New York, 1941.

More specifically, the square of z is

$$z^2 = r^2 e^{2i\theta}$$

another complex number having a modulus that is the square of the modulus of z and an argument that is twice the argument of z. The product of two complex numbers $z_1 = r_1 e^{i\theta_1}$ and $z_2 = r_2 e^{i\theta_2}$ is

$$z_1 z_2 = r_1 r_2 e^{i(\theta_1 + \theta_2)}$$

which is a complex number having a modulus that is the product of the moduli of z_1 and z_2 and an argument that is the sum of the arguments of z_1 and z_2. The process of multiplication of a complex number, say $re^{i\theta}$, by another complex number $Re^{i\varphi}$ may be thought of as an operation on $re^{i\theta}$ by $Re^{i\varphi}$ that consists of stretching (or shrinking) the modulus from r to Rr and of increasing the argument by the angle φ, changing the argument to $\theta + \varphi$. For example, $2e^{i\pi/4}$ when multiplied by $3e^{i\pi/2}$ stretches the modulus to 6 and rotates the segment $+90$ deg, so that its argument is $\pi/2 + \pi/4 = 135$ deg. This concept is of value in considering the mapping function discussed in the latter part of this chapter. Multiplication in cartesian form is

$$z_1 z_2 = (x_1 + iy_1)(x_2 + iy_2) = x_1 x_2 - y_1 y_2 + i(y_1 x_2 + x_1 y_2)$$

Division of one complex number by another is easily carried out when they are expressed in exponential form:

$$\frac{z_1}{z_2} = \frac{r_1 e^{i\theta_1}}{r_2 e^{i\theta_2}} = \frac{r_1}{r_2} e^{i(\theta_1 - \theta_2)}$$

This is an operation which is not defined in vector analysis. Division in cartesian form requires an extra step, thus

$$\frac{z_1}{z_2} = \frac{x_1 + iy_1}{x_2 + iy_2} = \frac{(x_1 + iy_1)(x_2 - iy_2)}{(x_2 + iy_2)(x_2 - iy_2)}$$

$$= \frac{(x_1 + iy_1)(x_2 - iy_2)}{x_2^2 + y_2^2}$$

$$= \frac{x_1 x_2 + y_1 y_2}{x_2^2 + y_2^2} + i \frac{y_1 x_2 - x_1 y_2}{x_2^2 + y_2^2}$$

The denominator is made real by multiplying both numerator and denominator by the denominator when i is replaced by $-i$ (its conjugate).

The complex number i may be expressed $e^{i\pi/2}$, as its modulus is 1 and its argument is $\pi/2$. Multiplication by i then may be written

$$iz = ire^{i\theta} = e^{i\pi/2} re^{i\theta} = re^{i(\theta + \pi/2)}$$

which shows clearly the rotation through $+90$ deg, caused by multiplication by i.

47. Conjugate Complex Numbers. The conjugate of a complex number is obtained from a complex number by replacing i by $-i$; thus

$$z = x + iy, \qquad \bar{z} = x - iy$$

where the bar over a complex number is used to indicate its conjugate. The product of any complex number and its conjugate is a real number:

$$z\bar{z} = (x + iy)(x - iy) = x^2 + y^2$$

or

$$z\bar{z} = re^{i\theta}re^{-i\theta} = r^2 = x^2 + y^2$$

Likewise the sum of a complex number and its conjugate is a real number, equal to twice the real part:

$$z + \bar{z} = x + iy + x - iy = 2x$$

The difference between a complex number and its conjugate is always a pure imaginary number

$$z - \bar{z} = x + iy - (x - iy) = i2y$$

48. The Logarithm of a Complex Number. The logarithm of a complex number is conveniently obtained from the exponential form; thus

$$\left.\begin{array}{l} z = x + iy = re^{i\theta} \\ \ln z = \ln r + i\theta \\ \qquad = \tfrac{1}{2}\ln(x^2 + y^2) + i \tan^{-1}\dfrac{y}{x} \end{array}\right\} \tag{3}$$

It is observed that the logarithm of a complex number is another complex number having its real part equal to the logarithm of the modulus and its imaginary part equal to the argument. As

$$e^{i(\theta + 2\pi)} = e^{i\theta}(\cos 2\pi + i \sin 2\pi) = e^{i\theta}$$

it is evident that the imaginary part of the logarithm is indeterminate to the extent of $2k\pi$, where k is any integer. This indeterminate feature of $\ln z$ may be avoided in many practical problems by restricting the range of θ, *e.g.*,

$$-\pi < \theta \leq \pi \qquad \text{or} \qquad 0 \leq \theta < 2\pi$$

49. Functions of a Complex Variable. Cauchy-Riemann Equations. When x and y in the complex number $z = x + iy$ are considered variables, then z is said to be a complex variable. Defining w as another complex variable such that

$$w = f(z) = f(x + iy)$$

w may be separated into its real part and its imaginary part, called ϕ and ψ, respectively,

$$w = \phi(x,y) + i\psi(x,y)$$

where ϕ and ψ are both real functions of x,y.

The function $f(z)$ is said to be a function of a complex variable if (1) within some region there is one and only one value of $f(z)$ for each value of z and that value is finite and (2) the function has a one-valued derivative at each point within the region. Within this region the function is said to be *holomorphic, regular,* or *analytic.*

Further consideration of (2) leads to relationships that must be fulfilled by a function if it is analytic. A complex derivative

$$\lim_{\delta z \to 0} \frac{f(z + \delta z) - f(z)}{\delta z}$$

may approach its limit in an infinite number of ways, *e.g.*, $\delta z = \delta x + i\,\delta y$ where δx may be related to δy by $\delta x = c\,\delta y$, with c an arbitrary constant. Condition (2) states that no matter which way the limit is taken, the derivative must have the same value. Two different paths by which the limit may be approached are considered, and conditions therefrom required to make the derivatives the same are necessary for existence of the one-valued derivative but are not sufficient. For the first path, δz is allowed to approach zero in the x-direction; *i.e.*, let $\delta y = 0$ first, then take the limit as δx approaches zero. This gives

$$\lim_{\substack{\delta y = 0 \\ \delta x = 0}} \frac{f(z + \delta z) - f(z)}{\delta x + i\,\delta y} = \lim_{\delta x = 0} \frac{f(z + \delta x) - f(z)}{\delta x} = \frac{\partial f}{\partial x}$$

where the last term comes from the second term which is the definition of a partial derivative (Sec. 5). For the second path, δz is allowed to approach zero in the y-direction by letting $\delta x = 0$ first; thus

$$\lim_{\substack{\delta x = 0 \\ \delta y = 0}} \frac{f(z + \delta z) - f(z)}{\delta x + i\,\delta y} = \frac{1}{i} \lim_{\delta y = 0} \frac{f(z + i\,\delta y) - f(z)}{\delta y} = \frac{1}{i} \frac{\partial f}{\partial y}$$

Since the derivative must be the same in either case if $f(z)$ is a function of a complex variable,

$$\frac{\partial f}{\partial x} = \frac{1}{i} \frac{\partial f}{\partial y} \tag{4}$$

However,

$$f(z) = w = \phi + i\psi$$

and therefore,

$$\frac{\partial f}{\partial x} = \frac{\partial \phi}{\partial x} + i\frac{\partial \psi}{\partial x}, \qquad \frac{\partial f}{\partial y} = \frac{\partial \phi}{\partial y} + i\frac{\partial \psi}{\partial y}$$

Substituting into Eq. (4),

$$\frac{\partial \phi}{\partial x} + i \frac{\partial \psi}{\partial x} = \frac{1}{i} \left(\frac{\partial \phi}{\partial y} + i \frac{\partial \psi}{\partial y} \right)$$

Equating the real parts on each side of the equation and then the imaginary parts,

$$\frac{\partial \phi}{\partial x} = \frac{\partial \psi}{\partial y}, \qquad \frac{\partial \phi}{\partial y} = - \frac{\partial \psi}{\partial x} \tag{5}$$

These relations are known as the Cauchy-Riemann equations. Books on analysis show that they are not only necessary but sufficient, provided the four partial derivatives $\frac{\partial \phi}{\partial x}, \frac{\partial \phi}{\partial y}, \frac{\partial \psi}{\partial x}, \frac{\partial \psi}{\partial y}$ are continuous.

Any $f(z) = f(x + iy)$ that is defined throughout a region and that has a derivative throughout the region satisfies the Cauchy-Riemann equations. Points on the plane where the Cauchy-Riemann equations are not satisfied are known as singular points.

50. Relation of Functions of a Complex Variable to Irrotational Flow. Differentiating the first of Eqs. (5) with respect to x and the second with respect to y and adding give

$$\frac{\partial^2 \phi}{\partial x^2} + \frac{\partial^2 \phi}{\partial y^2} = 0 \tag{6}$$

which is the Laplace equation in two-dimensional cartesian coordinates. Therefore, by considering ϕ to be a velocity potential, the real part of any function of a complex variable is a possible flow case, as it satisfies the continuity equation and is irrotational.

Similarly, differentiating the first of Eqs. (5) with respect to y and the second with respect to x and subtracting one from the other

$$\frac{\partial^2 \psi}{\partial x^2} + \frac{\partial^2 \psi}{\partial y^2} = 0 \tag{7}$$

showing that the pure imaginary part of any function of a complex variable may also be the velocity potential for a flow case. As the Cauchy-Riemann equations are identical to the relations developed at the end of Sec. 16 between stream function and velocity potential, it is obvious that when $\phi =$ constant are considered as equipotential lines, the lines $\psi =$ constant form an orthogonal system that are streamlines.

The proof that any $f(z) = f(x + iy)$ which is defined throughout a region and has a derivative throughout a region is a possible fluid flow case is as follows:

$$w = f(z) = f(x + iy) = \phi + i\psi, \qquad z = x + iy$$

To obtain the expression $\nabla^2 w$

$$\frac{\partial w}{\partial x} = \frac{dw}{dz}\frac{\partial z}{\partial x} = \frac{dw}{dz} \qquad \text{since} \qquad \frac{\partial z}{\partial x} = 1,$$

$$\frac{\partial^2 w}{\partial x^2} = \frac{d}{dz}\left(\frac{dw}{dz}\right)\frac{\partial z}{\partial x} = \frac{d^2 w}{dz^2}$$

and similarly,

$$\frac{\partial w}{\partial y} = \frac{dw}{dz}\frac{\partial z}{\partial y} = i\frac{dw}{dz} \qquad \text{since} \qquad \frac{\partial z}{\partial y} = i,$$

$$\frac{\partial^2 w}{\partial y^2} = \frac{d}{dz}\left(i\frac{dw}{dz}\right)\frac{\partial z}{\partial y} = -\frac{d^2 w}{dz^2}$$

Hence

$$\nabla^2 w = \frac{\partial^2 w}{\partial x^2} + \frac{\partial^2 w}{\partial y^2} = 0$$

and since

$$\nabla^2 w = \nabla^2(\phi + i\psi) = \nabla^2\phi + i\,\nabla^2\psi = 0$$
$$\nabla^2\phi = 0, \qquad \nabla^2\psi = 0$$

which are the conditions for ϕ and ψ to be either velocity potentials or stream functions for two-dimensional irrotational flow of an ideal fluid.

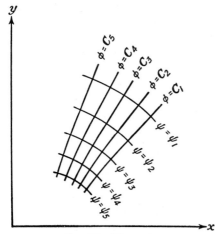

The functions ϕ, ψ are called *conjugate functions; i.e.*, the real part of an analytic function is said to be the conjugate of the imaginary part. The curves obtained by

$$\phi(x,y) = \text{constant},$$
$$\psi(x,y) = \text{constant}$$

form an orthogonal system in the xy-plane in that they intersect at right angles for every point in the plane where the function is regular. Allowing the constants c in

Fig. 42.—Flow net.

$$\phi = c, \qquad \psi = c$$

to take on values in arithmetic progression, the plotted curves in the xy-plane form an orthogonal network that, in the limit as the intervals in the arithmetical progression approach zero, is composed of infinitesimal squares. Such a system is referred to as a flow net, as shown in Fig. 42.

The simplest example of a flow net is given by

$$w = z = \phi + i\psi = x + iy$$

or

$$\phi = x, \qquad \psi = y$$

As ϕ and ψ take on constant values the resulting equipotential lines and streamlines are straight lines parallel to the y- and x-axes, respectively. A less simple example is

$$w = z^2 = (x + iy)^2 = x^2 - y^2 + i2xy$$

where now

$$\phi = x^2 - y^2, \qquad \psi = 2xy$$

the equipotential lines are hyperbolas with asymptotes $y = \pm x$, and the streamlines are rectangular hyperbolas with the coordinate axes as asymptotes, as in Fig. 45. Because both ϕ and ψ satisfy the Laplace equation, the streamlines may be taken as the equipotential lines and the equipotential lines as streamlines, thereby obtaining another flow case.

CONFORMAL MAPPING

51. Theory of Conformal Representation. The two complex variables w and z have been discussed in the preceding sections. They are given by

$$w = f(z) = \phi + i\psi, \qquad z = x + iy$$

where ϕ, ψ, x, y are real and

$$\phi = \phi(x,y), \qquad \psi = \psi(x,y)$$

A graph may be prepared in which ϕ is plotted as abscissa and ψ as ordinate in a manner similar to the plot of x and y. The $\phi\psi$-plane is called the w-plane and the xy-plane the z-plane. For each point in the z-plane there will correspond at least one point on the w-plane whose location is determined by the values of x,y and the functional relations $\phi(x,y)$, $\psi(x,y)$.

A curve or continuous line in the z-plane may be "transformed" or "mapped" onto the w-plane by determining corresponding points on the w-plane for points on the line in the z-plane. In this manner any configuration may be transformed from one plane into another configuration on the other plane, and the relation between the configurations will depend upon the functional relationship $w = f(z) = \phi + i\psi$. The transformation is said to be "conformal" if corresponding infinitesimal configurations in the two planes are similar.

An infinitesimal line element δz in the z-plane may be considered to be transformed into the corresponding infinitesimal line element δw in the w-plane by the operator $\dfrac{dw}{dz}$ as

$$\delta w = \frac{dw}{dz}\,\delta z \tag{8}$$

For infinitesimal figures to remain similar in the transformation they may be stretched or shrunk, but always by the same amount in every

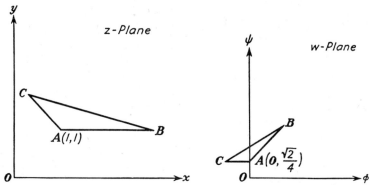

Fig. 43.—Conformal transformation of infinitesimal figure.

direction, and they may be rotated, but again every element must be rotated by the same amount. To fulfill these conditions $\dfrac{dw}{dz}$ must have one value only (although usually complex) at a point; *i.e.*, it must not change with orientation. This condition, however, is exactly that imposed by the Cauchy-Riemann equations. Thinking of $\dfrac{dw}{dz}$ as a constant complex number at any point, it then stretches and rotates δz as it is transformed into δw. This mapping process cannot be carried out at singular points, where $\dfrac{dw}{dz}$ becomes zero or infinite.

To illustrate the mapping process, the infinitesimal triangle ABC in the z-plane of Fig. 43 is mapped onto the w-plane, say by the function $w = \dfrac{\sqrt{2}\,z^2}{8}$. Let A be at $1 + i$ in the z-plane, which may be written $z = \sqrt{2}\,e^{i\pi/4}$. The point A is at

$$w = \frac{\sqrt{2}}{8}\,(\sqrt{2}\,e^{i\pi/4})^2 = \frac{\sqrt{2}}{4}\,e^{i\pi/2} = 0 + i\,\frac{\sqrt{2}}{4}$$

in the w-plane. At A,

$$\frac{dw}{dz} = z\,\frac{\sqrt{2}}{4}\Bigg]_{z=\sqrt{2}e^{i\pi/4}} = \frac{1}{2}e^{i\pi/4}$$

hence, multiplying δz by $\frac{1}{2}e^{i\pi/4}$ rotates it through $+45$ deg and decreases its corresponding length in the w-plane by one-half.   The lengths AB, AC are mapped over onto the w-plane in this manner.   Remembering that the figure is infinitesimal, BC is also rotated through $+45$ deg and shrunk one-half in length.   At other points than the immediate vicinity of A, $\dfrac{dw}{dz}$ takes on different values, so that large figures suffer a distortion.

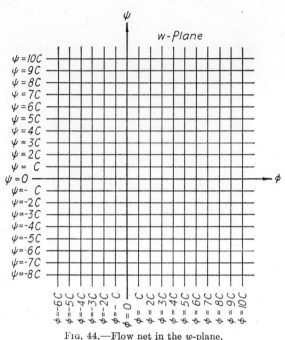

Fig. 44.—Flow net in the w-plane.

The simple function $w = Az$, where A is complex, say $a + ib$, has the same stretching and rotating value at all points, since

$$\frac{dw}{dz} = a + ib = \sqrt{a^2 + b^2}\, e^{i\,\tan^{-1}(b/a)}$$

is constant over the plane.   For this case finite figures also remain similar.

Many cases of simple transformations will be studied in the next chapter.

52. Application of Conformal Mapping to Ideal Fluid Flow. The real and pure imaginary parts of any function of $z = x + iy$ with continuous partial derivatives each provide a velocity potential or stream function for irrotational flow of an ideal fluid, as discussed in Sec. 50. Furthermore, any streamline in steady flow may be replaced by a solid boundary, as it satisfies the conditions for a boundary.

The w-plane (Fig. 44) is always considered to have a flow net that consists of a rectangular grid parallel to the ϕ- and ψ-axes. The vertical lines are equipotential lines. The flow represented by the w-plane is that of an infinite fluid flowing uniformly in the x-direction. Both families of lines may be mapped onto the z-plane by any analytic function of the form $w = f(z)$, and the lines $\phi = $ constant, $\psi = $ constant are still equipotential lines and streamlines in the z-plane. The particular configuration in the z-plane depends entirely upon $f(z)$, as the flow net in the w-plane is always the same. As $\psi_2 - \psi_1$, the difference in value of ψ between two streamlines (Sec. 16), represents the discharge, or flow, between those lines per unit of thickness, that same flow will prevail between corresponding lines in the z-plane. The spacing of the grid in the w-plane is not important, except that both ϕ and ψ should vary by the same increment from line to line. In illustrations of flow examples the w-plane is not usually shown, since it is the same in all cases.

The solution of two-dimensional fluid flow problems is attacked by the indirect method of investigating various functions to determine the shape of boundaries to which they might apply.

53. Inverse Transformations. In some important cases the function relating w to z cannot be solved explicitly for w. When it is solved in the form $z = f(w)$, the function is said to be an inverse function. Expanding,

$$z = f(w) = f(\phi + i\psi) = x(\phi,\psi) + iy(\phi,\psi)$$

Many useful flow nets are obtained by the inverse transformation. It gives an entirely different flow net than $w = f(z)$, where f is the same in both cases.

54. Complex Potential.   Complex Velocity. The complex variable w, when defined by

$$w = \phi + i\psi$$

where ϕ and ψ satisfy the Cauchy-Riemann equations, is called the *complex potential.* ϕ is the velocity potential and ψ the stream function.

Taking the partial derivative of the complex potential with respect to x,

$$\frac{\partial w}{\partial x} = \frac{\partial \phi}{\partial x} + i\frac{\partial \psi}{\partial x} = \frac{dw}{dz}\frac{dz}{\partial x} = \frac{dw}{dz} \tag{9}$$

since from

$$z = x + iy, \qquad \frac{\partial z}{\partial x} = 1$$

Introducing u and v, the velocity components,

$$u = -\frac{\partial \phi}{\partial x}, \qquad v = \frac{\partial \psi}{\partial x}$$

and substituting in Eq. (9), a useful relation is obtained:

$$\frac{dw}{dz} = -u + iv \tag{10}$$

where $\frac{dw}{dz}$ is called the *complex velocity*. The real part of $\frac{dw}{dz}$ is the negative of the velocity component in the x-direction, and the pure imaginary part is the velocity component in the y-direction. The complex velocity usually provides the most convenient means of determining velocity at a point. The speed, or magnitude of the velocity, is given by

$$\left|\frac{dw}{dz}\right| = \sqrt{u^2 + v^2}$$

or by means of the conjugate complex velocity

$$\frac{d\bar{w}}{d\bar{z}} = -u - iv$$

$$\frac{dw}{dz}\frac{d\bar{w}}{d\bar{z}} = (-u + iv)(-u - iv) = u^2 + v^2$$

For example, to find the velocity at $z = 1 + 2i$ from the function $w = z^2 + z$,

$$\frac{dw}{dz} = 2z + 1 = 2(1 + 2i) + 1 = 3 + 4i$$

Hence,

$$u = -3, \qquad v = 4, \qquad q = \sqrt{u^2 + v^2} = 5$$

Stagnation points are those points in the flow where the velocity is zero. They are given by

$$\frac{dw}{dz} = 0$$

as both real and pure imaginary parts of a complex number must vanish when it equals zero.

Some transformations are more conveniently carried out by using an intermediate plane, say from w to ζ, then from ζ to z; *e.g.*,

$$w = \sin \zeta, \qquad z = \ln \zeta$$

where the second relation is an inverse transformation. The complex velocity is given by

$$\frac{dw}{dz} = \frac{dw}{d\zeta}\frac{d\zeta}{dz} = \frac{dw}{d\zeta}\frac{1}{dz/d\zeta} = \zeta \cos \zeta$$

The derivative $\frac{dw}{dz}$ has been given two interpretations, one as the complex velocity and the other as an operator by means of which elements of the z-plane are transformed into elements in the w-plane. When $\dfrac{dw}{dz}$ is zero or infinite, the transformation is not conformal. Those points are singular points and should be excluded from the flow net by drawing small circles around them.

In the following paragraphs some relationships are developed that are useful in dealing with inverse transformations. Starting with $z = f(w)$, the derivative becomes

$$- \frac{dz}{dw} = - \frac{1}{dw/dz} = \frac{1}{u - iv} = \frac{1}{q}\left(\frac{u}{q} + i\frac{v}{q}\right) \tag{11}$$

where $q = \sqrt{u^2 + v^2}$, and the last term is obtained by multiplying numerator and denominator of the third term by $u + iv$. Defining ζ by

$$\zeta = - \frac{dz}{dw}$$

it is a complex number that has the direction (or argument) of the velocity vector at each point and a modulus that is the reciprocal of the speed $1/q$. u/q and v/q are direction cosines of the velocity vector, so that $(u/q + iv/q)$ is a unit velocity vector at each point throughout the flow.

From the inverse relations

$$x = x(\phi,\psi), \qquad y = y(\phi,\psi)$$

where

$$w = \phi + i\psi, \qquad z = x + iy$$

the following relations are developed:

$$\frac{\partial z}{\partial \phi} = \frac{\partial}{\partial \phi}(x + iy) = \frac{\partial x}{\partial \phi} + i\frac{\partial y}{\partial \phi} = \frac{dz}{dw}\frac{\partial w}{\partial \phi} = \frac{dz}{dw}$$

and

$$\frac{\partial z}{\partial \psi} = \frac{\partial}{\partial \psi}(x + iy) = \frac{\partial x}{\partial \psi} + i\frac{\partial y}{\partial \psi} = \frac{dz}{dw}\frac{\partial w}{\partial \psi} = i\frac{dz}{dw}$$

Equating $\dfrac{dz}{dw}$ in the last two expressions,

$$\frac{\partial x}{\partial \phi} + i\frac{\partial y}{\partial \phi} = \frac{1}{i}\left(\frac{\partial x}{\partial \psi} + i\frac{\partial y}{\partial \psi}\right)$$

from which

$$\frac{\partial x}{\partial \phi} = \frac{\partial y}{\partial \psi}, \qquad \frac{\partial y}{\partial \phi} = - \frac{\partial x}{\partial \psi} \tag{12}$$

These relations take the place of the Cauchy-Riemann equations.

Since $\dfrac{1}{q}$ is the modulus of $\dfrac{dz}{dw}$, and since $\dfrac{dz}{dw} = \dfrac{\partial z}{\partial \phi}$,

$$\frac{1}{q^2} = \left(\frac{\partial x}{\partial \phi}\right)^2 + \left(\frac{\partial y}{\partial \phi}\right)^2 \tag{13}$$

From the interrelationships this may be written

$$\frac{1}{q^2} = \left(\frac{\partial x}{\partial \phi}\right)^2 + \left(\frac{\partial x}{\partial \psi}\right)^2 = \left(\frac{\partial y}{\partial \psi}\right)^2 + \left(\frac{\partial y}{\partial \phi}\right)^2$$

$$= \left(\frac{\partial x}{\partial \psi}\right)^2 + \left(\frac{\partial y}{\partial \psi}\right)^2$$

$$= \frac{\partial x}{\partial \phi}\frac{\partial y}{\partial \psi} - \frac{\partial x}{\partial \psi}\frac{\partial y}{\partial \phi}$$

Exercises

1. Find the complex numbers given by

(a) $(2 - 3i) + (4 + i)$
(b) $(1 + i)(6 - 3i)$
(c) $(1 - i)(2 + i)(1 - 3i)$
(d) $\dfrac{1 + 2i}{1 - i}$
(e) $\ln (3 + 4i)$
(f) $\ln (0 + i)$
(g) $\ln (-1)$

2. Given $z^6 = -i$, by use of the DeMoivre's theorem find six roots of z. Plot them on Argand diagram.

HINT: Solve $z^6 = e^{i[(3\pi/2)+2\pi k]}$ for z, using $k = 0, 1, 2, 3, 4, 5$

3. Separate the following functions of z into their real and imaginary parts ϕ and ψ.

(a) $\dfrac{1}{z}$

(b) $\dfrac{z}{\bar{z}}$

(c) $\ln z^2$

(d) $\dfrac{z + \bar{z}}{z - \bar{z}}$

(e) e^{iz}

(f) $\dfrac{1}{z} + \bar{z}$

4. Show that the Cauchy-Riemann equations in polar coordinates are

$$\frac{\partial \phi}{\partial r} = \frac{\partial \psi}{r\,\partial \theta}, \quad \frac{\partial \phi}{r\,\partial \theta} = -\frac{\partial \psi}{\partial r}.$$

5. Which of the following functions are functions of a complex variable?

(a) $\sqrt{r} \cos \dfrac{\theta}{2} + i\sqrt{r} \sin \dfrac{\theta}{2}$

(b) $\dfrac{1}{x^2} + i\dfrac{1}{y^2}$

(c) $\dfrac{x}{x^2 + y^2} - i\dfrac{y}{x^2 + y^2}$

(d) $x^2 - y^2 + x - i(2xy + y)$

6. For those functions in Exercise 5 which satisfy the Cauchy-Riemann equations, determine the function of z, by substituting $x = z - iy$ or $r = ze^{-i\theta}$ and simplifying.

7. Map the triangle having vertices at $(1,0)$, $(2,1)$, $(0,0)$ in the z-plane onto the w-plane using the following functions:

    (a) $w = (1 + i)z$ (b) $w = z^2$

8. Plot the velocity vector at the vertices of the triangle in the z-plane of Exercise 7.

9. Determine the stagnation points for the flow nets given by:

    (a) $w = \ln z + z$ (b) $z = \ln w$

    (c) $w = \dfrac{i}{z} + z$ (d) $w = z^2 + iz$

10. Determine the velocity at $(1,1)$ and $(3,0)$ in the z-plane for the inverse transformations

    (a) $z = w^2$ (b) $z = \dfrac{i}{w}$

    (c) $z = e^{iw}$

CHAPTER VI

TWO-DIMENSIONAL FLOW EXAMPLES

As the theory of complex variables leads to an infinite number of solutions of the Laplace equation, yielding streamlines as well as equipotential lines, all examples in two-dimensional flow are treated from the standpoint of conformal transformations. Sufficient examples are given in this chapter to provide an understanding of the manipulations required to analyze simple functions. Flow nets are constructed in most cases as an aid in selecting the proper function for particular problems. The flow cases have been classified according to the type of function or the type of coordinate system employed. Flow around circular cylinders is studied in Chap. VII as the initial transformations involved in determining flow around the Kutta-Joukowsky airfoil. Chapter VIII deals with a special type of two-dimensional flow having free boundaries, called *free streamlines*.

SIMPLE CONFORMAL TRANSFORMATIONS

55. Exponential Transformations. w = Az^n. By specifying whether A is real, complex, or pure imaginary and by specifying n, many different transformations can be obtained. Several special values of A and n are examined.

n = 1. A Complex. Let $A = a + ib$; then

$$w = (a + ib)(x + iy) = \phi + i\psi = ax - by + i(bx + ay),$$

$$\frac{dw}{dz} = A = a + ib$$

Since $\dfrac{dw}{dz}$ is a constant, finite figures as well as infinitesimal figures will be similar in the w- and z-planes. Writing $A = Re^{i\theta}$,

$$\delta z = \frac{dz}{dw}\, \delta w = \frac{1}{R}\, e^{-i\theta}\, \delta w$$

showing that elements in the w-plane are elongated in the ratio $1/R : 1$ and rotated through $-\theta$ when mapped onto the z-plane. Hence, the rectangular ϕ, ψ-grid in the w-plane remains a rectangular grid in the z-plane. When A is real, $\theta = 0$, and scale only is changed in the two grids. When A is pure imaginary, the grid is rotated through ± 90 deg on the z-plane. There are no singular points on the finite plane.

n = 2. A Real. The equations pertinent to this case are

$$w = Az^2 = A(x + iy)^2 = A(x^2 - y^2) + iA2xy$$

and

$$\phi = A(x^2 - y^2), \qquad \psi = 2Axy, \qquad \frac{dw}{dz} = 2Az$$

As $\dfrac{dw}{dz}$ is a function of z, it varies in modulus and argument, distorting

FIG. 45.—Flow net for $w = Az^2$.

the rectangular $\phi\psi$-grid in the w-plane into the flow net shown in Fig. 45. The vertical lines $\phi =$ constant in the w-plane become the family of hyperbolas $x^2 - y^2 =$ constant having axes coincident with the x-axis and asymptotes $y = \pm x$. The horizontal lines $\psi =$ constant in the w-plane map into the rectangular hyperbolas $xy =$ constant, having axes $y = \pm x$ and with the coordinate axes as asymptotes.

Expressing the stream function in polar coordinates,

$$\psi = 2Ar \cos\theta\, r \sin\theta = Ar^2 \sin 2\theta$$

the streamline $\psi = 0$ is given by $\theta = 0$, $\theta = \pi/2$, or the positive coordinate axes. This transformation may be taken to be the fluid motion adjacent to two perpendicular walls. At the origin $\dfrac{dw}{dz} = 0$; therefore, it is a singular point that must be excluded from the conformal map.

When A is pure imaginary, the streamlines and equipotential lines of Fig. 45 are interchanged.

n = −1. A Real. Hence, as $w = A/z$,

$$w = \frac{A}{x + iy} = \frac{A(x - iy)}{(x + iy)(x - iy)}$$

$$= \frac{Ax}{x^2 + y^2} - i\frac{Ay}{x^2 + y^2}$$

from which

$$\phi = \frac{Ax}{x^2 + y^2}, \qquad \psi = \frac{-Ay}{x^2 + y^2}, \qquad \frac{dw}{dz} = -\frac{A}{z^2}$$

The equipotential lines $\phi = C$ are circles through the origin with

FIG. 46.—Flow net for $w = A/z$.

centers on the x-axis in the z-plane, readily seen when the expression for ϕ is placed in standard form for a circle in cartesian coordinates; thus

$$\left(x - \frac{A}{2\phi}\right)^2 + y^2 = \frac{A^2}{4\phi^2}$$

In a similar manner the streamlines $\psi = $ constant are circles through the origin with centers on the y-axis, *viz.*,

$$x^2 + \left(y + \frac{A}{2\psi}\right)^2 = \frac{A^2}{4\psi^2}$$

The flow net is given in Fig. 46. At the origin $\dfrac{dw}{dz}$ becomes infinite; therefore, it is a singular point not included in the mapping process. This is
the case of a doublet of strength A at the origin; it is discussed more
generally in Sec. 61.

Fig. 47.—Flow net $w = A/z^2$, a double family of lemniscates.

n = −2. A Real. Using polar coordinates, $z = re^{i\theta}$,

$$w = \frac{A}{z^2} = \frac{A}{r^2} e^{-i2\theta} = \frac{A}{r^2} (\cos 2\theta - i \sin 2\theta)$$

from which

$$\phi = \frac{A}{r^2} \cos 2\theta, \qquad \psi = -\frac{A}{r^2} \sin 2\theta, \qquad \frac{dw}{dz} = -\frac{2A}{z^3}$$

The equipotential lines and streamlines $\phi = C$, $\psi = C$ are the double
family of lemniscates shown in Fig. 47. The origin is a singular point
that must be excluded from the region.

n = π/α. A Real. This is a general case where α may take any
value between 0 and 2π. Using polar coordinates,

$$w = Az^{\pi/\alpha} = Ar^{\pi/\alpha}\left(\cos \frac{\pi}{\alpha} \theta + i \sin \frac{\pi}{\alpha} \theta\right)$$

from which

$$\phi = Ar^{\pi/\alpha} \cos \frac{\pi\theta}{\alpha}, \qquad \psi = Ar^{\pi/\alpha} \sin \frac{\pi\theta}{\alpha}, \qquad \frac{dw}{dz} = \frac{\pi}{\alpha} Az^{(\pi/\alpha)-1}$$

The streamline $\psi = 0$ is given by $\theta = 0$ and $\theta = \alpha$. Hence, the uniform flow of the w-plane is transformed into flow between two plane boundaries at an angle α. At the origin $\dfrac{dw}{dz}$ becomes zero if α is less than π and becomes infinite if α is greater than π. In either case it is a singular point that must be excluded from the conformal region. Figure 48 shows the flow net for two values of α: 45 and 225 deg.

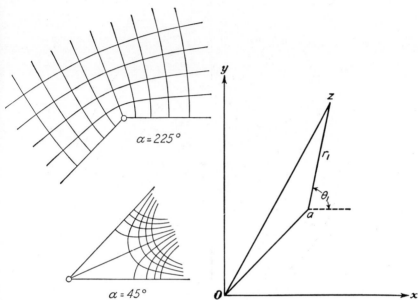

$\alpha = 225°$

$\alpha = 45°$

FIG. 48.—Flow along two inclined plane surfaces.

FIG. 49.—Special coordinates for source at $z = a$.

56. Source. Vortex. $w = -A\mu \ln (z - a)$.

The case where A is the real number unity is considered first, with μ a positive real constant and a complex. Plotting a in Fig. 49 the complex number $z - a$, obtained by vector subtraction of $\overline{oa}$ from $\overline{oz}$, has a modulus r_1 and an argument θ_1,

$$z - a = r_1 e^{i\theta_1}$$

Substituting into the complex potential,

$$w = -\mu \ln r_1 e^{i\theta_1} = -\mu \ln r_1 - i\mu\theta_1$$

and

$$\phi = -\mu \ln r_1, \qquad \psi = -\mu\theta_1, \qquad \frac{dw}{dz} = -\frac{\mu}{z - a}$$

The equipotential lines are concentric circles about a, and the streamlines are radial lines through a, as shown in Fig. 50.   The point a is a singular point that is excluded from the region.   The velocity vector is everywhere in the direction of radial lines through a, as $\dfrac{\partial \phi}{\partial \theta} = 0$.   The velocity a distance r_1 from a is

$$-\frac{\partial \phi}{\partial r_1} = \frac{\mu}{r_1}$$

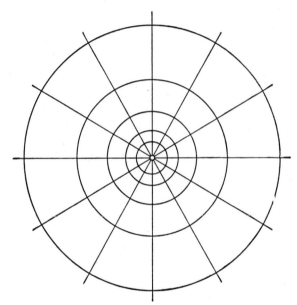

FIG. 50.—Flow net for source or vortex.

and the discharge or flow from point a (per unit thickness) is

$$2\pi r_1 \frac{\mu}{r_1} = 2\pi\mu$$

hence, this is a source of strength $2\pi\mu$ located at $z = a$.   Letting A be -1 results in a sink instead of a source.

Considering A imaginary, say $-i$, then

$$w = i\mu \ln (z - a) = i\mu \ln r_1 - \mu\theta_1$$

from which

$$\phi = -\mu\theta_1, \qquad \psi = \mu \ln r_1$$

Figure 50 is the flow pattern for this case also, except that the equipotential lines are now the radial lines through a and the streamlines the concentric circles around a.   The line integral of the velocity about any closed path including the point a is $2\pi\mu$. Selecting a streamline as path, the magnitude of the velocity is constant at μ/r_1, it is in the direction of the path; hence, the line integral (Sec. 23) gives

$$2\pi r_1 \frac{\mu}{r_1} = 2\pi\mu$$

which is the circulation (Sec. 25). This is the case of a vortex of strength, $\kappa = 2\pi\mu$, at a.

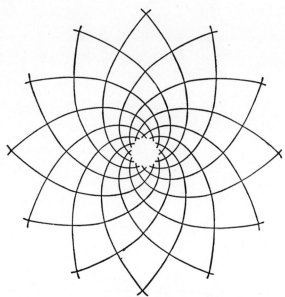

Fig. 51.—Combination of source and vortex.

When A is complex, say $1 - i$,

$$\phi = -\mu\theta_1 - \mu \ln r_1$$
$$\psi = -\mu\theta_1 + \mu \ln r_1$$

the flow net is obtained from that of Fig. 50. The streamlines and equipotential lines are drawn for the same increments $\Delta\phi = \Delta\psi$. Then to construct $\phi = $ constant in going from one radial line to the next in the positive direction, ϕ decreases by $\Delta\phi$ along a circular path, but by dropping to the next smaller circular path ϕ has increased by $\Delta\phi$. The procedure then is to connect diagonal vertices of the polygons in Fig. 50. Similarly, for the streamlines the opposite diagonal vertices are connected. Figure 51 shows the new flow pattern.

57. Hyperbolic Functions and Identities. In many of the two-dimensional flow cases hyperbolic functions are required. They will be defined in this section, and their relations to trigonometric functions given. Many useful relationships are listed, which can be worked out from the definitions.

There are six hyperbolic functions that are comparable to the trigonometric functions. They are designated by adding the letter h to the trigonometric abbreviations: sinh x, cosh x, tanh x, coth x, sech x, csch x.

The basic definitions may be stated in terms of exponentials:

$$\sinh x = \frac{1}{\operatorname{csch} x} = \frac{e^x - e^{-x}}{2}$$

$$\cosh x = \frac{1}{\operatorname{sech} x} = \frac{e^x + e^{-x}}{2}$$

$$\tanh x = \frac{1}{\operatorname{coth} x} = \frac{e^x - e^{-x}}{e^x + e^{-x}}$$

The corresponding definitions for trigonometric functions, obtained from DeMoivre's theorem (Sec. 46), are

$$\sin x = \frac{1}{\csc x} = \frac{e^{ix} - e^{-ix}}{2i}$$

$$\cos x = \frac{1}{\sec x} = \frac{e^{ix} + e^{-ix}}{2}$$

$$\tan x = \frac{1}{\cot x} = -i\frac{e^{ix} - e^{-ix}}{e^{ix} + e^{-ix}}$$

From the above definitions, the following relationships can be established:

$$\sinh ix = i \sin x \tag{1}$$

$$\cosh ix = \cos x \tag{2}$$

$$\sinh x = -i \sin ix \tag{3}$$

$$\cosh x = \cos ix \tag{4}$$

$$\sinh (-x) = -\sinh x \tag{5}$$

$$\cosh (-x) = \cosh x \tag{6}$$

$$\cosh^2 x - \sinh^2 x = 1 \tag{7}$$

$$1 - \tanh^2 x = \operatorname{sech}^2 x \tag{8}$$

$$1 - \coth^2 x = -\operatorname{csch}^2 x \tag{9}$$

$$\sinh (x \pm y) = \sinh x \cosh y \pm \cosh x \sinh y \tag{10}$$

$$\cosh (x \pm y) = \cosh x \cosh y \pm \sinh x \sinh y \tag{11}$$

$$\tanh (x \pm y) = \frac{\tanh x \pm \tanh y}{1 \pm \tanh x \tanh y} \tag{12}$$

$$\sinh (2x) = 2 \sinh x \cosh x \tag{13}$$

$$\cosh (2x) = \cosh^2 x + \sinh^2 x$$

$$= 2 \cosh^2 x - 1$$

$$= 1 + 2 \sinh^2 x \tag{14}$$

$$\tanh(2x) = \frac{2\tanh x}{1 + \tanh^2 x} \tag{15}$$

$$\sinh\left(\frac{x}{2}\right) = \sqrt{\frac{1}{2}(\cosh x - 1)} \tag{16}$$

$$\cosh\left(\frac{x}{2}\right) = \sqrt{\frac{1}{2}(\cosh x + 1)} \tag{17}$$

$$\tanh\left(\frac{x}{2}\right) = \frac{\cosh x - 1}{\sinh x} = \frac{\sinh x}{\cosh x + 1} \tag{18}$$

$$d\sinh x = \cosh x \, dx \tag{19}$$

$$d\cosh x = \sinh x \, dx \tag{20}$$

$$d\tanh x = \operatorname{sech}^2 x \, dx \tag{21}$$

$$d\coth x = -\operatorname{csch}^2 x \, dx \tag{22}$$

$$d\operatorname{sech} x = -\operatorname{sech} x \tanh x \, dx \tag{23}$$

$$d\operatorname{csch} x = -\operatorname{csch} x \coth x \, dx \tag{24}$$

$$\sinh z = \sinh(x + iy)$$
$$= \sinh x \cos y + i \cosh x \sin y \tag{25}$$

$$\cosh z = \cosh(x + iy)$$
$$= \cosh x \cos y + i \sinh x \sin y \tag{26}$$

$$\sin z = \sin x \cosh y + i \cos x \sinh y \tag{27}$$

$$\cos z = \cos x \cosh y - i \sin x \sinh y \tag{28}$$

The inverse hyperbolic functions exist and are expressible in terms of logarithms as follows:

$$y = \sinh^{-1} x, \qquad x = \sinh y = \frac{e^{2y} - 1}{2e^y} \tag{29}$$

or

$$e^{2y} - 2xe^y - 1 = 0, \qquad e^y = x + \sqrt{x^2 + 1}$$

As e^y is always positive, the positive sign must be used before the radical. Taking the logarithm

$$y = \sinh^{-1} x = \ln(x + \sqrt{x^2 + 1}) \qquad \text{any } x \tag{30}$$

Similarly,

$$\cosh^{-1} x = \ln(x \pm \sqrt{x^2 - 1}) \qquad x > 1 \tag{31}$$

$$\tanh^{-1} x = \frac{1}{2} \ln \frac{1 + x}{1 - x} \qquad x^2 < 1 \tag{32}$$

$$\coth^{-1} x = \frac{1}{2} \ln \frac{x + 1}{x - 1} \qquad x^2 > 1 \tag{33}$$

$$\operatorname{sech}^{-1} x = \ln\left(\frac{1}{x} \pm \sqrt{\frac{1}{x^2} - 1}\right) \qquad x < 1 \tag{34}$$

$$\operatorname{csch}^{-1} x = \ln\left(\frac{1}{x} + \sqrt{\frac{1}{x^2} + 1}\right) \qquad \text{any } x \tag{35}$$

The differentials are

$$d \sinh^{-1} x = \frac{dx}{\sqrt{x^2 + 1}} \tag{36}$$

$$d \cosh^{-1} x = \frac{\pm dx}{\sqrt{x^2 - 1}} \tag{37}$$

$$d \tanh^{-1} x = \frac{dx}{1 - x^2} = d \coth^{-1} x \tag{38}$$

$$d \operatorname{sech}^{-1} x = \frac{-dx}{x\sqrt{1 - x^2}} \tag{39}$$

$$d \operatorname{csch}^{-1} x = \frac{-dx}{x\sqrt{1 + x^2}} \tag{40}$$

58. Source and Sink of Equal Strength. The complex potential for a source and sink of equal strength $2\pi\mu$, located for simplicity at $(a,0)$ and $(-a,0)$, respectively, is

$$w = -\mu \ln (z - a) + \mu \ln (z + a)$$

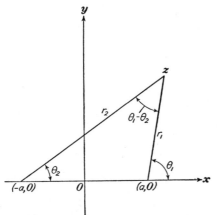

Using the notation of Fig. 52, where

$$z - a = r_1 e^{i\theta_1}, \qquad z + a = r_2 e^{i\theta_2}$$

for any point z, then

$$w = -\mu \ln \frac{z - a}{z + a}$$

$$= -\mu \ln \frac{r_1}{r_2} - i\mu(\theta_1 - \theta_2)$$

and

$$\phi = -\mu \ln \frac{r_1}{r_2}$$

$$\psi = -\mu(\theta_1 - \theta_2)$$

$$\frac{dw}{dz} = \frac{-2\mu a}{(z - a)(z + a)}$$

Fig. 52.—Notation for source and sink.

Expressing $r_1{}^2 = (x - a)^2 + y^2$, $r_2{}^2 = (x + a)^2 + y^2$, the equation for ϕ may take the form

$$\left(x - a \coth \frac{\phi}{\mu}\right)^2 + y^2 = \left(a \operatorname{csch} \frac{\phi}{\mu}\right)^2$$

For equipotential lines, ϕ = constant; this is the equation for a family of circles of radius $a \operatorname{csch} \phi/\mu$ with centers at $(a \coth \phi/\mu,\ 0)$. Likewise, by expressing $\theta_1 = \tan^{-1}[y/(x - a)]$, $\theta_2 = \tan^{-1}[y/(x + a)]$, the equation for ψ becomes

$$x^2 + \left(y + a \cot \frac{\psi}{\mu}\right)^2 = \left(a \csc \frac{\psi}{\mu}\right)^2$$

Hence, the streamlines are circles of radius $a \csc \psi/\mu$ having centers at $(0, -a \cot \psi/\mu)$. The flow pattern is given in Fig. 53.

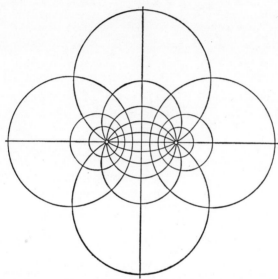

FIG. 53.—Flow pattern for source and sink of equal strength or for two equal vortices of opposite sign.

Multiplying the complex potential by $-i$, a new complex potential is obtained in which the streamlines of Fig. 53 become equipotential lines, and vice versa. This is the flow pattern due to two vortices of equal strength but different signs at $(a,0)$ and $(-a,0)$. The circulation around any path enclosing $(a,0)$ and excluding $(-a,0)$ has a circulation $2\pi\mu$, while the circulation around any path embracing $(-a,0)$ and not $(a,0)$ is $-2\pi\mu$; for a circuit enclosing both singular points the circulation is zero. The points $(a,0)$, $(-a,0)$ must be excluded from the conformal region.

59. Flow around a Circular Cylinder Due to an External Source. The combination of two equal sources at Q and P and a sink of the same strength at O (Fig. 54) produces a flow pattern with a closed circle of radius $\overline{OR} = \sqrt{\overline{OQ} \cdot \overline{OP}}$. Since a closed streamline may be replaced by a solid boundary, the resulting flow is that due to a source near a circular cylinder. The complex potential is

$$w = \mu \ln z - \mu \ln (z - \overline{OQ}) - \mu \ln (z - \overline{OP})$$

Substituting

$$z = r_1 e^{i\theta_1}, \qquad z - \overline{OQ} = r_2 e^{i\theta_2}, \qquad z - \overline{OP} = r_3 e^{i\theta_3}$$

$$w = \mu \ln \frac{r_1 e^{i\theta_1}}{r_2 e^{i\theta_2} r_3 e^{i\theta_3}} = \mu \ln \frac{r_1}{r_2 r_3} + i\mu(\theta_1 - \theta_2 - \theta_3)$$

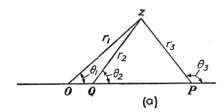

(a)

and

$$\phi = \mu \ln \frac{r_1}{r_2 r_3},$$

$$\psi = \mu(\theta_1 - \theta_2 - \theta_3)$$

$$\frac{dw}{dz} = \mu \frac{\overline{OR}^2 - z^2}{z(z - \overline{OP})(z - \overline{OQ})}$$

Since $\overline{OR} = \sqrt{\overline{OP} \cdot \overline{OQ}}$, Q is the inverse of P in the circle. Then, as in Sec. 38,

Angle ORQ = angle RPO

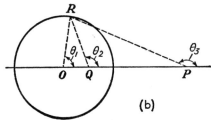

(b)

FIG. 54.—Sources at Q and P and sink at O.

by similar triangles. Writing ψ for any point R on the circle, from Fig. 54b,

$$\psi = -\mu(\theta_3 + \theta_2 - \theta_1) = -\mu(\theta_3 + \text{angle } ORQ)$$
$$= -\mu(\theta_3 + \text{angle } RPO) = -\mu\pi$$

Hence, $\psi = -\mu\pi$ is satisfied by any point R on the circle of radius $\overline{OR} = \sqrt{\overline{OQ} \cdot \overline{OP}}$. The source at P is the only singular point outside the cylinder. The flow pattern is shown in Fig. 55.

60. Flow Due to a Series of Equal and Equidistant Sources along the y-axis. The complex potential

$$w = C \ln \sinh \frac{\pi z}{a}$$

may be written

$$w = C \ln \left[\frac{\pi z}{a} \prod_{n=1}^{\infty} \left(1 + \frac{z^2}{n^2 a^2} \right) \right]$$

since[1]

$$\sinh t = t \prod_{n=1}^{\infty} \left(1 + \frac{t^2}{n^2\pi^2}\right)$$

where Π means the product of terms. Expanding the products and taking the logarithm,

$$w = C \ln \frac{\pi z}{a} + C \ln \left(1 + \frac{z^2}{a^2}\right) + C \ln \left(1 + \frac{z^2}{4a^2}\right)$$
$$+ C \ln \left(1 + \frac{z^2}{9a^2}\right) + \cdots$$

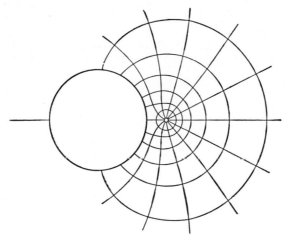

Fig. 55.—Flow net for source near a circular cylinder.

or expanding further,

$$w = C[\ln z + \ln (z - ia) + \ln (z + ia) + \ln (z - 2ia) + \ln (z + 2ia)$$
$$+ \cdots] + C \left(\ln \frac{\pi}{a} - \ln a^2 - \ln 4a^2 \cdots \right)$$

In this form the complex potential is seen to be that due to an infinite series of sources,[2] all of strength $2\pi C$, located at the points $(0,0)$, $(0, \pm a)$, $(0, \pm 2a)$ .... The constant part of the series can be dropped, as it contributes nothing to the flow pattern; the velocity potential and stream function are always subject to addition of an arbitrary constant.

[1] Smithsonian Mathematical Formulae and Tables of Elliptic Functions, p. 130, The Smithsonian Institute, Washington, 1922.

[2] If C is negative, the series is composed of sources, if positive, the series is composed of sinks.

Returning to the first form of the complex potential, the real and pure imaginary parts may be separated by substituting $z = x + iy$ and expanding, using Eq. (25). Then

$$\phi = \tfrac{1}{2}C \ln \tfrac{1}{2}\left(\cosh \frac{2\pi x}{a} - \cos \frac{2\pi y}{a}\right)$$

$$\psi = C \tan^{-1} \frac{\tan \pi y/a}{\tanh \pi x/a}$$

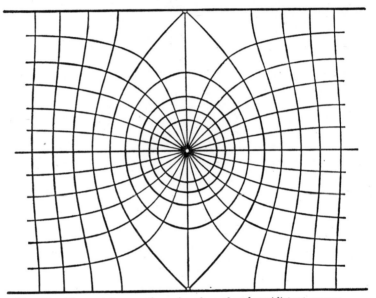

Fig. 56.—Flow net for one of a series of equal and equidistant sources.

From the symmetry of the problem there will be no flow across the lines

$$y = 0, \ \pm \frac{a}{2}, \ \pm a, \ \pm \frac{3a}{2}, \ \pm 2a, \ \pm \frac{5a}{2}, \ \cdots$$

The formulas may apply to a source midway between two fixed parallel boundaries. The flow pattern is shown in Fig. 56.

61. Doublets. The two-dimensional doublet of strength μ, located at $z = z_0$ and with the axis making an angle α with the x-axis, has the complex potential

$$w = \mu \frac{e^{i\alpha}}{z - z_0}$$

The flow pattern is that of Fig. 46 when the origin there is placed at $z = z_0$ and the flow net is rotated through the angle α.

The complex potential may be derived by applying the limiting process to a source and a sink, as required by the definition of a doublet (Sec. 22). Consider a source at $z_0 + a$ of strength $2\pi m$ and a sink of the same strength at $z_0 - a$ (Fig. 57). The complex potential, as in Sec. 58, is

$$w = -m \ln [z - (z_0 + a)] + m \ln [z - (z_0 - a)]$$

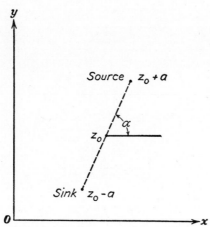

FIG. 57.—Notation for source and sink.

To apply the limiting process as the complex number a approaches zero, let

$$a = a_1 e^{i\alpha}$$

where a_1 is the absolute value of a or its modulus and α is its argument. Rewriting the complex potential,

$$w = -m \ln \left[(z - z_0) \left(1 - \frac{a_1 e^{i\alpha}}{z - z_0} \right) \right] + m \ln \left[(z - z_0) \left(1 + \frac{a_1 e^{i\alpha}}{z - z_0} \right) \right]$$

$$= -m \ln \left(1 - \frac{a_1 e^{i\alpha}}{z - z_0} \right) + m \ln \left(1 + \frac{a_1 e^{i\alpha}}{z - z_0} \right)$$

Using the infinite series expansions,

$$\ln (1 - t) = -t - \frac{t^2}{2} - \frac{t^3}{3} - \cdots$$

$$\ln (1 + t) = t - \frac{t^2}{2} + \frac{t^3}{3} - \cdots$$

when a_1 is small

$$w = -m \left[-\frac{a_1 e^{i\alpha}}{z - z_0} - \frac{a_1^2 e^{i2\alpha}}{2(z - z_0)^2} - \frac{a_1^3 e^{i3\alpha}}{3(z - z_0)^3} \cdots \right]$$

$$+ m \left[\frac{a_1 e^{i\alpha}}{z - z_0} - \frac{a_1^2 e^{i2\alpha}}{2(z - z_0)^2} + \frac{a_1^3 e^{i3\alpha}}{3(z - z_0)^3} \cdots \right]$$

$$= 2a_1 m \left[\frac{e^{i\alpha}}{z - z_0} + \frac{a_1^2 e^{i3\alpha}}{3(z - z_0)^3} \cdots \right]$$

From the definition of the doublet, the product of the strength and distance between source and sink is to remain constant in the limit as the distance approaches zero. The formula is in the form suitable for taking the limit. Let $2a_1 m = \mu$; then as a_1 approaches zero,

$$w = \frac{\mu e^{i\alpha}}{z - z_0}$$

which is the general equation of the two-dimensional doublet. The special case of a doublet at the origin with axis in the positive x-direction

$$w = \frac{\mu}{z}$$

has been treated in Sec. 55.

62. Infinite Number of Doublets along the y-axis. The complex potential for a doublet with axis in the positive x-direction may be obtained by differentiation of the complex potential for a sink. For example, the sink of strength $2\pi\mu$ at $z = a$, from Sec. 58, is given by

$$w = \mu \ln (z - a)$$

Differentiating with respect to z, the new complex potential is

$$w = \frac{\mu}{z - a}$$

which is a doublet of strength μ, as in Sec. 61.

Similarly, differentiation of the complex potential for a series of equal and equidistant sinks along the y-axis (Sec. 60) produces the complex potential for a series of equal and equidistant doublets along the y-axis, with axes in the positive x-direction. Term by term differentiation of the series expansion for series of sinks proves this to be the case. The velocity potential and stream function can be obtained as follows: First, the complex potential of Sec. 60 is differentiated

$$w = \frac{d}{dz} \left(C \ln \sinh \frac{\pi z}{a} \right) = \frac{\pi C}{a} \coth \frac{\pi z}{a}$$

$$= C' \coth \frac{\pi z}{a}$$

The doublets are located at $(0,0)$, $(0,\pm a)$, $(0,\pm 2a)$, ... Substituting $z = x + iy$ and separating w into its real and pure imaginary parts ϕ, ψ, by using Eqs. (25) and (26),

$$\phi = C' \frac{\sinh 2\pi x/a}{\cosh 2\pi x/a - \cos 2\pi y/a}$$

$$\psi = -C' \frac{\sin 2\pi y/a}{\cosh 2\pi x/a - \cos 2\pi y/a}$$

from which the equipotential and streamlines of Fig. 58 are plotted.

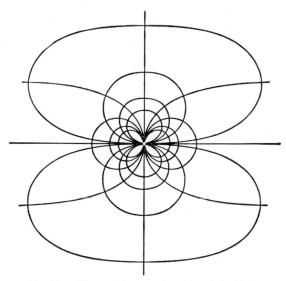

Fig. 58.—Flow net for one of a series of doublets.

Superposing upon the foregoing flow pattern a uniform flow of velocity U in the negative x-direction,

$$w = Uz, \qquad \phi = Ux, \qquad \psi = Uy$$

the complex potential becomes

$$w = Uz + C' \coth \frac{\pi z}{a}$$

The velocity potential and stream function are

$$\phi = Ux + C' \frac{\sinh 2\pi x/a}{\cosh 2\pi x/a - \cos 2\pi y/a}$$

$$\psi = Uy - C' \frac{\sin 2\pi y/a}{\cosh 2\pi x/a - \cos 2\pi y/a}$$

The equation for the streamline $\psi = 0$ may be expressed

$$y\left(\cosh\frac{2\pi x}{a} - \cos\frac{2\pi y}{a}\right) = \frac{C'}{U}\sin\frac{2\pi y}{a} = \frac{C'}{U}\left[\frac{2\pi y}{a} - \left(\frac{2\pi y}{a}\right)^3\frac{1}{3!} + \cdots\right]$$

using the series expansion for $\sin 2\pi y/a$. The equation is satisfied by $y = 0$. Factoring y out, the streamline $\psi = 0$ is also satisfied by

$$\cosh\frac{2\pi x}{a} - \cos\frac{2\pi y}{a} = \frac{2\pi C'}{aU} - \left(\frac{2\pi}{a}\right)^3\frac{C'}{U}\frac{y^2}{3!} + \cdots$$

This is the equation of an oval with origin as center. Its semidiameter along the x-axis is obtained by setting $y = 0$,

$$\cosh\frac{2\pi x_0}{a} - 1 = \frac{2\pi C'}{aU}$$

The semidiameter y_0 along the y-axis is obtained by setting $x = 0$,

$$1 - \cos\frac{2\pi y_0}{a} = \frac{C'}{Uy_0}\sin\frac{2\pi y_0}{a}$$

Expressing this in terms of the angle $\pi y_0/a$,

$$y_0\tan\frac{\pi y_0}{a} = \frac{C'}{U}$$

By setting $C' = \pi b^2 U/a$, where b is small compared with a, then as

$$\sinh\frac{\pi x_0}{a} \cong \frac{\pi x_0}{a}, \qquad \tan\frac{\pi y_0}{a} \cong \frac{\pi y_0}{a}$$

for small values of $\pi x_0/a$ and $\pi y_0/a$, substitution shows that both semi-diameters are approximately equal to b.

Since any closed streamline can be replaced by a solid boundary, ϕ and ψ can be construed as velocity potential and stream function for flow through a grating of parallel cylindrical bars of circular cross section. For small values of x and y the stream function takes the form

$$\psi = Uy\left(1 - \frac{b^2}{x^2 + y^2}\right)$$

when $C' = \pi b^2 U/a$. Letting $a = \sqrt{4\pi}\,b$ and solving for semidiameters,

$$x = 0.254a, \qquad y = 0.25a$$

showing the section is very close to circular when the bars take up half the spacing between centers. Figure 59 shows the flow net. Lines of symmetry are the y-axis and the lines parallel to the x-axis through $(0,0)$, $(0,\pm a/2)$, $(0,\pm a)$, $(0,\pm 3a/2)$. . . .

INVERSE TRANSFORMATIONS

63. Elliptic Coordinates. z = c cosh w. This function when solved explicitly for z, is an inverse of the function $w = f(z)$. When used for

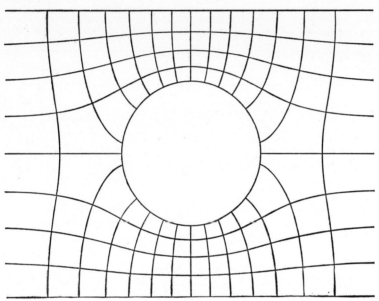

Fig. 59.—Flow net for steady flow of an infinite fluid around a cylinder between parallel walls.

mapping it is referred to as an inverse transformation. The relationships peculiar to inverse transformations have been worked out in Secs. 53 and 54. The rectangular grid in the w-plane is the same as for simple transformations, and the new grid in the z-plane is sought. $\dfrac{dz}{dw}$ must be single-valued and not zero for the transformation to be conformal.

Separating the function into its real and pure imaginary parts, using Eq. (26),

$$x + iy = c \cosh (\phi + i\psi) = c \cosh \phi \cos \psi + ic \sinh \phi \sin \psi$$

from which

$$\left. \begin{array}{l} x = c \cosh \phi \cos \psi \\ y = c \sinh \phi \sin \psi \end{array} \right\} \tag{41}$$

Eliminating ψ,

$$\frac{x^2}{c^2 \cosh^2 \phi} + \frac{y^2}{c^2 \sinh^2 \phi} = 1 \qquad (42)$$

This shows that the equipotential lines $\phi = $ constant must appear in the z-plane as ellipses, having foci at (c,O) and $(-c,O)$.

Eliminating ϕ,

$$\frac{x^2}{c^2 \cos^2 \psi} - \frac{y^2}{c^2 \sin^2 \psi} = 1 \qquad (43)$$

yields a family of confocal hyperbolas for $\psi = $ constant, having the same foci as the family of ellipses.

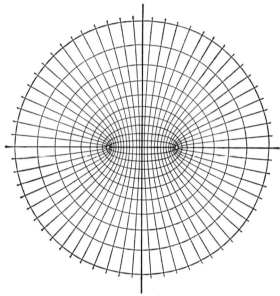

FIG. 60.—Elliptic coordinates. Flow through a rectangular slot.

The resulting grid in the z-plane (Fig. 60) may be referred to as elliptic coordinates. They necessarily form an orthogonal system, except at the singular points given by

$$\frac{dz}{dw} = c \sinh w$$

being either zero or infinite. The singular points are at the foci.

Taking the lines $\phi = $ constant for equipotential lines, this may be considered as the flow through an aperture of width $2c$ in the thin plate comprising all the x-axis except from $x = -c$ to $x = +c$. Taking the ellipses as streamlines, circulation around an elliptic cylinder is portrayed or, in the limit, circulation around the rectangular lamina extending from $x = -c$ to $x = +c$ on the x-axis.

64. Flow into a Rectangular Channel. z = w + e^w. The advantages of the special relations developed for inverse transformations are apparent from this function which cannot be solved explicitly for w. Separating into its real and pure imaginary parts,

$$x + iy = \phi + i\psi + e^\phi \cos \psi + ie^\phi \sin \psi$$

from which

$$x = \phi + e^\phi \cos \psi, \qquad y = \psi + e^\phi \sin \psi, \qquad \frac{dz}{dw} = 1 + e^w$$

The streamline $\psi = 0$ is given by

$$x = \phi + e^\phi, \qquad y = 0$$

As ϕ varies from plus infinity to minus infinity, x also varies from plus infinity to minus infinity; hence, the x-axis is the streamline $\psi = 0$. The streamline $\psi = \pi$ is given by

$$x = \phi - e^\phi, \qquad y = \pi$$

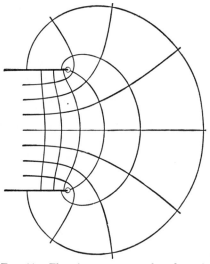

As ϕ varies from plus infinity to zero, x varies from minus infinity to minus 1. As ϕ varies from minus infinity to zero, x again varies from minus infinity to minus 1. Therefore, the line

$$y = \pi, \qquad -\infty < x \leq -1$$

Fig. 61.—Flow into a rectangular channel.

is the streamline $\psi = \pi$. This line may be considered as bent back upon itself. Similarly the streamline $\psi = -\pi$ is

$$y = -\pi, \qquad -\infty < x \leq -1$$

The streamlines lying between $\psi = -\pi$ and $\psi = \pi$ are all contained within the two straight lines $y = \pm\pi$, $-\infty < x \leq -1$ for negative values of ϕ. For positive values of ϕ the intermediate streamlines fan out and cover the complete z-plane. The flow net is shown in Fig. 61. It represents the flow into or out of a canal bounded by two parallel walls.

The singular points are given by

$$+ 1 + e^\phi \cos \psi + ie^\phi \sin \psi = 0$$

which is satisfied by

$$\phi = 0, \qquad \psi = \pm\pi$$

These are the points $x = -1$, $y = \pm\pi$, or the end of the canal walls.

65. Flow into Channel with Diverging Walls. The preceding example may be generalized in such a manner that the flow into a channel with diverging walls is obtained. Starting with the function

$$z = \frac{1 - \beta/\pi}{\beta/\pi} \left(1 - e^{-\beta w/\pi}\right) + e^{(1-\beta/\pi)w}$$

it can be reduced to

$$z = w + e^w$$

by applying the limiting process as β approaches zero. Separating into real and pure imaginary parts and solving for x and y,

$$x = \frac{1 - \beta/\pi}{\beta/\pi} \left(1 - e^{-\beta\phi/\pi} \cos \frac{\beta\psi}{\pi}\right) + e^{(1-\beta/\pi)\phi} \cos \left[\left(1 - \frac{\beta}{\pi}\right)\psi\right]$$

$$y = \frac{1 - \beta/\pi}{\beta/\pi} e^{-\beta\phi/\pi} \sin \frac{\beta\psi}{\pi} + e^{(1-\beta/\pi)\phi} \sin \left[\left(1 - \frac{\beta}{\pi}\right)\psi\right]$$

The streamline $\psi = 0$ is the x-axis, for $\beta < \pi$. For $\psi = \pm\pi$ the expression for x and y becomes

$$x = \frac{1 - \beta/\pi}{\beta/\pi} - \cos \beta \left[\frac{1 - \beta/\pi}{\beta/\pi} e^{-\beta\phi/\pi} + e^{(1-\beta/\pi)\phi}\right]$$

$$y = \pm \sin \beta \left[\frac{1 - \beta/\pi}{\beta/\pi} e^{-\beta\phi/\pi} + e^{(1-\beta/\pi)\phi}\right]$$

Eliminating ϕ in these equations,

$$y = \mp x \tan \beta \pm \frac{1 - \beta/\pi}{\beta/\pi} \tan \beta$$

As ϕ varies from minus infinity to plus infinity, x remains less than or equal to

$$\frac{\pi}{\beta} (1 - \cos \beta) - 1$$

Hence, the streamlines $\psi = \pm\pi$ are straight lines making an angle β with the x-axis, extending from minus infinity to

$$x = \frac{\pi}{\beta} (1 - \cos \beta) - 1$$

which again approaches -1 as β approaches zero.   Changing the sign of w changes the direction of flow.   The flow net is shown in Fig. 62, for $\beta = 30$ deg.

Singular points occur at the ends of the channel, $\phi = 0$, $\psi = \pm\pi$ where the velocity is infinite.

66. Idealized Flow around Two-dimensional Pitot Tube.   By superposing uniform unit velocity in the positive x-direction on the flow into a channel with parallel walls (Sec. 64), the net flow into the channel is reduced to zero.   This is similar to the flow around an idealized Pitot tube except that being two-dimensional, the parallel walls take the place of a cylindrical tube.

Unit velocity in the x-direction is given by

$$w'' = -z,$$
$$\phi'' = -x,$$
$$\psi'' = -y$$

where the primes are used to distinguish this function from the final complex potential that is sought.   From Sec. 64 the complex potential for flow into the channel is given by

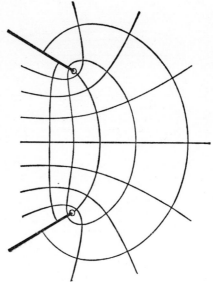

FIG. 62.—Flow into channel with diverging walls.

$$z = w' + e^{w'}, \qquad w' = \phi' + i\psi' \tag{44}$$

or

$$x = \phi' + e^{\phi'}\cos\psi', \qquad y = \psi' + e^{\phi'}\sin\psi' \tag{45}$$

The resultant complex potential is

$$w = \phi + i\psi = w' + w'' = \phi' + i\psi' + \phi'' + i\psi''$$
$$= \phi' - x + i(\psi' - y)$$

or

$$\phi' = \phi + x, \qquad \psi' = \psi + y, \qquad w' = w + z$$

Substituting these relations in Eqs. (44) and (45),

$$w + e^{w+z} = 0 \tag{46}$$
$$\phi + e^{\phi+x}\cos(\psi + y) = 0 \tag{47}$$
$$\psi + e^{\phi+x}\sin(\psi + y) = 0 \tag{48}$$

Eliminating first y and then x from Eqs. (47) and (48) produces the desired equations

$$x = -\phi + \ln \sqrt{\phi^2 + \psi^2} \tag{49}$$

$$y = -\psi + \tan^{-1} \frac{\psi}{\phi} \tag{50}$$

Combining these two expressions,

$$z = -w + \ln w \tag{51}$$

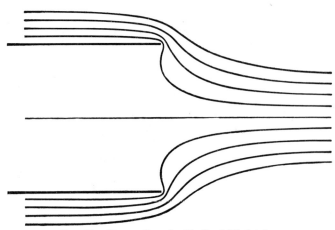

Fig. 63.—Stream lines for idealized Pitot tube.

is the inverse function. This may be obtained directly from Eq. (46), since

$$\ln (-w) = \ln w + \text{constant}$$

and the constant does not affect the flow pattern.

From Eqs. (49) and (50) it is evident that x is an even function of ψ and y an odd function of ψ. The x-axis is an axis of symmetry. When $\psi = 0$, from Eq. (48)

$$e^{\phi+x} \sin y = 0$$

which is satisfied by $y = 0, \pm\pi$. From Eq. (47)

$$\phi = -e^{\phi+x} \cos y$$

hence, if ϕ is negative, $y = 0$; and if ϕ is positive, $y = \pm\pi$. From Eq. (49) when ϕ varies from minus infinity to zero, x varies from plus infinity to minus infinity; but when ϕ is positive, x varies from minus infinity to minus 1. Therefore, the streamline

$$\psi = 0, \qquad y = \pm\pi \qquad -\infty < x \leq -1$$

may be taken for the walls of the Pitot tube.

To show that the velocity is zero an infinite distance into the tube, by use of Sec. 54 and Eq. (51)

$$-\frac{dz}{dw} = \frac{1}{q}\left(\frac{u}{q} + i\frac{v}{q}\right) = 1 - \frac{1}{w}$$

when $y = 0$, $\psi = 0$; hence,

$$\frac{1}{u} = 1 - \frac{1}{\phi}$$

or

$$u = \frac{\phi}{\phi - 1}$$

As ϕ approaches zero, x approaches minus infinity and u approaches zero. A few streamlines are portrayed in Fig. 63.

67. Boundary Conditions for the Translation of Any Cylinder through an Infinite Fluid. The boundary condition for translation of a cylinder through an infinite fluid is simply expressed in terms of the stream function. Let the cylinder have the velocity U in the x-direction, as in Fig. 64. The boundary condition must state that the velocity of fluid normal to the surface at the boundary equals the velocity of the surface normal to itself. Taking s positive in the direction shown and the normal positive when directed into the fluid, δs, δn form a right-handed system similar to δx, δy. The fluid velocity in the n-direction is given

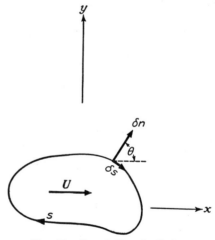

Fig. 64.—Translation of cylinder.

$\frac{\partial \psi}{\partial s}$. The velocity of the boundary normal to itself is $U \cos \theta$. However, $\cos \theta$ may be expressed as $-\frac{dy}{ds}$. Hence, the boundary condition, in differential form, is

$$\frac{\partial \psi}{\partial s} = -U\frac{dy}{ds}$$

Integrating around the periphery,

$$\psi = -Uy + \text{constant} \tag{52}$$

where the constant is arbitrary.   Any function ψ that satisfies this equation for a closed cylinder provides the stream flow pattern for values of $\psi = $ constant.   Two examples of the use of this equation are given.
   The function

$$\psi = -Uy$$

satisfies the boundary condition [Eq. (52)] identically for any shape of body.   It is the case of uniform fluid motion with constant velocity $u = U$.   It must then be a case of fluid contained within a cylinder that is in translation.   This is the only possible irrotational motion which can be given to a fluid by translation of a cylindrical shell if the region is simply connected.[1]   The complex potential is

$$w = -Uz$$

   Expressing Eq. (52) in polar coordinates,

$$\psi = -Ur \sin \theta + \text{constant}$$

Examining the function

$$\psi = -\frac{AU}{r} \sin \theta$$

the boundary condition is satisfied by substituting into it and solving for A; thus

$$-\frac{AU}{r} \sin \theta = -Ur \sin \theta$$

letting the constant be zero.   This equation is satisfied if

$$r = \sqrt{A} = a$$

Hence,

$$\psi = -\frac{Ua^2}{r} \sin \theta$$

is the stream function for translation of a circular cylinder through an infinite fluid otherwise at rest.   The complex potential is

$$w = \frac{Ua^2}{z}$$

The flow net is given by Fig. 46 if the circle is drawn in with center at the origin.
   68. Translation of an Elliptic Cylinder.   Elliptic coordinates were defined in Sec. 63.   By making use of two transformations, first from w to an auxiliary plane, the ζ-plane, then from the ζ-plane to the z-plane

[1] A region is simply connected if any closed curve within the region can be shrunk to zero size without leaving the region.

using the elliptic coordinate transformation, several interesting flow patterns are determined.

Consider the relations

$$w = Ce^{-\zeta}, \qquad z = c \cosh \zeta \qquad \zeta = \xi + i\eta \qquad (53)$$

where the center relation is for elliptic coordinates and the last gives ξ and η as the real and pure imaginary portions of ζ. C and c are real constants. The potential and stream functions are

$$\phi = Ce^{-\xi} \cos \eta, \qquad \psi = -Ce^{-\xi} \sin \eta \qquad (54)$$

Substituting ψ into the boundary condition [Eq. (52)],

$$-Ce^{-\xi} \sin \eta = -Uy + \text{constant}$$
$$= -Uc \sinh \xi \sin \eta + \text{constant}$$

from Eq. (41), Sec. 63. U is the velocity in the x-direction. Letting the arbitrary constant be zero, the boundary condition is satisfied by one value of ξ, say ξ_0, determined by

$$Ce^{-\xi_0} = Uc \sinh \xi_0$$

$\xi = \xi_0$ is the equation of an elliptic cylinder, with semimajor and semiminor axes a, b, respectively.

$$a = c \cosh \xi_0, \qquad b = c \sinh \xi_0, \qquad c = \sqrt{a^2 - b^2}$$

Solving for C,

$$C = e^{\xi_0} Uc \sinh \xi_0 = \sqrt{\frac{a + b}{a - b}}\, Ub$$

since

$$e^{\xi_0} = \sqrt{\frac{a + b}{a - b}}$$

from the definitions of the hyperbolic functions. Substituting C back into Eqs. (54),

$$\psi = -Ub \sqrt{\frac{a + b}{a - b}}\, e^{-\xi} \sin \eta$$

is the stream function for an elliptic cylinder of semiaxes a, b, moving parallel to the major axis with velocity U in an infinite fluid otherwise at rest. a is greater than or equal to b, as $\cosh \xi_0$ is never smaller than $\sinh \xi_0$.

The stream function for motion parallel to the minor axis may be obtained by considering the body translating in the positive y-direction with velocity V. The boundary condition is

$$\psi = Vx + \text{constant}$$

and the stream function is

$$\psi = Va \sqrt{\frac{a+b}{a-b}} \, e^{-\xi} \cos \eta$$

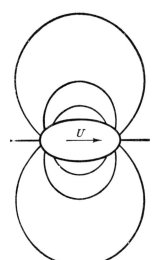

The complex potential is

$$w = iVa \sqrt{\frac{a+b}{a-b}} \, e^{-\xi} \qquad (55)$$

The two stream functions may be added to give the stream function for arbitrary direction of translation through proper choice of U and V.

Figures 65 and 66 show the unsteady streamlines for translation of an elliptic cylinder parallel to its major and minor axes, respectively.

When b is set equal to zero in Eq. (55), the flow case is for motion of a plate at right angles to its surface through an infinite fluid. The same streamline pattern is obtained as for elliptic cylinders. Singular points occur at the edges of the plate, where the velocity becomes infinite. The pattern, therefore,

Fig. 65.—Unsteady streamlines for translation of elliptic cylinder parallel to major axis.

differs greatly from that obtained by moving a thin plate through an actual fluid.

69. Kinetic Energy. In Sec. 14 the equation for kinetic energy of fluid was found to be

$$T = -\frac{\rho}{2} \int \phi \frac{\partial \phi}{\partial n} \, dS$$

with δn the normal to the boundary surface, drawn positive into the

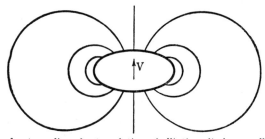

Fig. 66.—Unsteady streamlines for translation of elliptic cylinder parallel to minor axis.

fluid, and dS an element of the boundary surface. Reducing this to the two-dimensional case, with unit thickness considered,

$$\frac{\partial \phi}{\partial n} = \frac{\partial \psi}{\partial s}$$

with $\delta n, \delta s$ forming a right-handed system similar to x,y as in Fig. 67. The surface area element is now ds, and the equation becomes

$$T = -\frac{\rho}{2} \int \phi \frac{\partial \psi}{\partial s} \, ds = -\frac{\rho}{2} \int \phi \, d\psi \qquad (56)$$

which is evaluated over the boundary.

As an example of its use the kinetic energy of fluid due to translation of an elliptic cylinder is computed. The stream function, from Sec. 68, is

$$\psi = Va \sqrt{\frac{a+b}{a-b}} \, e^{-\xi} \cos \eta$$

and the corresponding potential function

$$\left(\text{since } \frac{\partial \phi}{\partial \xi} = \frac{\partial \psi}{\partial \eta}, \frac{\partial \phi}{\partial \eta} = -\frac{\partial \psi}{\partial \xi} \right)$$

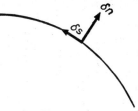

$$\phi = Va \sqrt{\frac{a+b}{a-b}} \, e^{-\xi} \sin \eta$$

FIG. 67.—Normal and surface elements.

Hence, the kinetic energy T is

$$T = +\frac{\rho}{2} V^2 a^2 \left(\frac{a+b}{a-b} \right) e^{-2\xi_0} \int_0^{2\pi} \sin^2 \eta \, d\eta$$

$$= \frac{\pi \rho}{2} V^2 a^2$$

70. Scale Factors for Two-dimensional Coordinate Systems. Let ξ, η represent any orthogonal two-dimensional coordinate system, related to x,y by

$$x = x(\xi,\eta)$$
$$y = y(\xi,\eta)$$

The condition for the system to be orthogonal is first determined. The slope of the curve $\eta = \eta_0$ is

$$\frac{dy}{dx} = \frac{dy(\xi,\eta_0)}{dx(\xi,\eta_0)} = \frac{\partial y/\partial \xi}{\partial x/\partial \xi}$$

Similarly the slope of the curve $\xi = \xi_0$ is

$$\frac{dy}{dx} = \frac{\partial y/\partial \eta}{\partial x/\partial \eta}$$

When these two slopes are negative reciprocals, the two curves intersect at right angles; therefore,

$$\frac{\partial y}{\partial \xi} \frac{\partial y}{\partial \eta} + \frac{\partial x}{\partial \xi} \frac{\partial x}{\partial \eta} = 0 \tag{57}$$

is the condition for orthogonality that must be satisfied at all points throughout the plane except singular points.

Since the velocity in any direction, say the s-direction, is given by $-\frac{\partial \phi}{\partial s}$, it is necessary to compute δs in terms of any orthogonal coordinate system ξ, η in order to determine the velocity conveniently at any point. This can be accomplished as follows:

$$ds^2 = dx^2 + dy^2, \qquad dx = \frac{\partial x}{\partial \xi} d\xi + \frac{\partial x}{\partial \eta} d\eta$$

$$dy = \frac{\partial y}{\partial \xi} d\xi + \frac{\partial y}{\partial \eta} d\eta$$

$$x = x(\xi, \eta), \qquad y = y(\xi, \eta)$$

Then

$$ds^2 = \left[\left(\frac{\partial x}{\partial \xi}\right)^2 + \left(\frac{\partial y}{\partial \xi}\right)^2 \right] d\xi^2 + 2 \left[\frac{\partial x}{\partial \xi} \frac{\partial x}{\partial \eta} + \frac{\partial y}{\partial \xi} \frac{\partial y}{\partial \eta} \right] d\xi\, d\eta$$

$$+ \left[\left(\frac{\partial x}{\partial \eta}\right)^2 + \left(\frac{\partial y}{\partial \eta}\right)^2 \right] d\eta^2$$

$$= E\, d\xi^2 + 2F\, d\xi\, d\eta + G\, d\eta^2$$

where E, F, G are Gauss numbers. F is zero for an orthogonal system [Eq. (57)]. Hence,

$$ds^2 = E\, d\xi^2 + G\, d\eta^2$$

Along the curve $\eta = \eta_0$, $d\eta = 0$ and

$$ds = \sqrt{E}\, d\xi$$

Similarly, along the curve $\xi = \xi_0$,

$$ds = \sqrt{G}\, d\eta$$

The quantities $\sqrt{E}$, $\sqrt{G}$ are scale factors. For conformal transformations the scale at any point must be the same in all directions; therefore,

$$\sqrt{E} = \sqrt{G}$$

The scale factor for any two-dimensional coordinate system that can be expressed as a function of a complex variable $z = z(\zeta)$ is given by

$$\sqrt{E} = \sqrt{\left(\frac{\partial x}{\partial \xi}\right)^2 + \left(\frac{\partial y}{\partial \xi}\right)^2} \tag{58}$$

The velocity components in the positive $\xi\eta$-directions are

$$v_\xi = -\frac{\partial\phi}{\sqrt{E}\,\partial\xi}, \qquad v_\eta = -\frac{\partial\phi}{\sqrt{E}\,\partial\eta}$$

when ϕ is expressed as a function of ξ,η.

For elliptic coordinates

$$x = c \cosh \xi \cos \eta$$
$$y = c \sinh \xi \sin \eta$$

and

$$\sqrt{G} = \sqrt{E} = c\sqrt{\sinh^2 \xi \cos^2 \eta + \cosh^2 \xi \sin^2 \eta}$$

71. Steady Flow around an Elliptic Cylinder. By superposing a uniform velocity U in the negative x-direction upon the flow system given by Eqs. (53), steady flow around an elliptic cylinder is obtained. The complex potential for uniform velocity $-U$ is

$$w = Uz = Uc \cosh \zeta$$

as $z = c \cosh \zeta$ is the elliptic coordinate relation. The new complex potential then becomes

$$w = Ub\sqrt{\frac{a+b}{a-b}}\,e^{-\zeta} + Uc \cosh \zeta \qquad (59)$$

from which the potential and stream functions may be written:

$$\phi = Ub\sqrt{\frac{a+b}{a-b}}\,e^{-\xi} \cos \eta + U\sqrt{a^2 - b^2}\cosh \xi \cos \eta \qquad (60)$$

and

$$\psi = -Ub\sqrt{\frac{a+b}{a-b}}\,e^{-\xi} \sin \eta + U\sqrt{a^2 - b^2}\sinh \xi \sin \eta \qquad (61)$$

Equation (61) gives $\psi = 0$ for $\eta = 0,\pi$, which is the x-axis except for $-c < x < +c$. $\psi = 0$ is also given by

$$\xi = \xi_0 = \tfrac{1}{2}\ln\frac{a+b}{a-b}$$

readily verified by substitution. Hence, the elliptic cylinder $\xi = \xi_0$ may be taken as a boundary.

Similarly, by superposing $w = -iVz$ upon the flow given by Eq. (55), the steady flow about an elliptic cylinder "broadside" to the flow is obtained; thus

$$w = iVa\sqrt{\frac{a+b}{a-b}}\,e^{-\zeta} - iVc \cosh \zeta \qquad (62)$$

from which

$$\phi = Va\sqrt{\frac{a+b}{a-b}}\,e^{-\xi}\sin\eta + V\sqrt{a^2-b^2}\sinh\xi\sin\eta \qquad (63)$$

and

$$\psi = Va\sqrt{\frac{a+b}{a-b}}\,e^{-\xi}\cos\eta - V\sqrt{a^2-b^2}\cosh\xi\cos\eta \qquad (64)$$

Equation (64) gives $\psi = 0$ for $\eta = \pi/2$, which is the y-axis. For

$$\xi = \xi_0 = \tfrac{1}{2}\ln\frac{a+b}{a-b},$$

ψ is also zero, which is the same elliptic cylinder as given by Eq. (61).

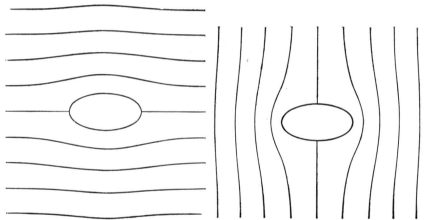

Fig. 68.—Steady flow around an elliptic cylinder $V = 0$.

Fig. 69.—Steady flow around an elliptic cylinder $U = 0$.

Equations (59) and (62) may be added to produce the case for steady flow about an elliptic cylinder with any arbitrary direction of the undisturbed velocity by proper selection of U and V. Figure 68 shows a few streamlines for the case where $V = 0$, Fig. 69 for the case where $U = 0$, and Fig. 70 for the case where $U = V$.

Setting $b = 0$ reduces the equations to flow around a rectangular lamina.

72. Boundary Conditions for Rotation of Any Cylinder in an Infinite Fluid. The boundary condition for rotation of a rigid cylinder about an axis through the origin is obtained from the original definition: The velocity component at the boundary normal to the boundary must equal the velocity of the boundary normal to itself. In Fig. 71 let δs denote an element of the surface of the cylinder. The positive direction of s

is as shown, and the positive normal to the surface is drawn into the
fluid.   The fluid velocity normal to the surface is $\dfrac{\partial \psi}{\partial s}$.   The velocity of a
point P on the surface is ωr, and the velocity normal to the surface is
$\omega r \cos \theta$.   From Fig. 71 $\cos \theta = \dfrac{dr}{ds}$, hence, the differential equation for
boundary condition is

$$\frac{\partial \psi}{\partial s} = \omega r \frac{dr}{ds}$$

which applies equally well to external or internal boundaries.

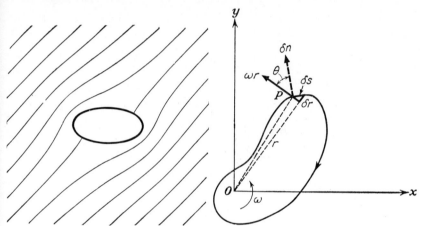

FIG. 70.—Steady flow around an elliptic   FIG. 71.—Rotation of cylinder about origin.
cylinder $U = V$.

Integrating,

$$\psi = \tfrac{1}{2}\omega r^2 + \text{constant} \qquad (65)$$

where the constant is arbitrary.   Some examples of rotating boundaries
are discussed in the following sections.

73. Fluid Contained within a Rotating Elliptic Cylinder.   The complex potential

$$w = iAz^2 \qquad (66)$$

has the potential and stream functions

$$\phi = -2Axy, \qquad \psi = A(x^2 - y^2)$$

Substituting ψ into the boundary condition [Eq. (65)], expressed in
cartesian coordinates,

$$A(x^2 - y^2) = \tfrac{1}{2}\omega(x^2 + y^2) - C$$

Rearranging,

$$x^2(\tfrac{1}{2}\omega - A) + y^2(\tfrac{1}{2}\omega + A) = C$$

This is the equation for a conic. Writing it in the standard form for an ellipse,

$$\frac{x^2}{\dfrac{C}{\tfrac{1}{2}\omega - A}} + \frac{y^2}{\dfrac{C}{\tfrac{1}{2}\omega + A}} = 1$$

which is an ellipse when $A < \tfrac{1}{2}\omega$. Denoting the semimajor and semi-

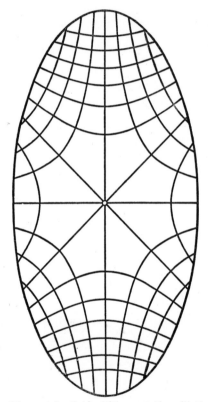

Fig. 72.—Flow net for fluid within a rotating elliptic cylinder.

minor axes by a, b, respectively,

$$a^2 = \frac{C}{\tfrac{1}{2}\omega - A}, \qquad b^2 = \frac{C}{\tfrac{1}{2}\omega + A}$$

Solving for A by eliminating C,

$$A = \tfrac{1}{2}\omega \frac{a^2 - b^2}{a^2 + b^2}$$

The stream function is

$$\psi = \tfrac{1}{2}\omega \frac{a^2 - b^2}{a^2 + b^2}\,(x^2 - y^2)$$

Since the fluid velocity approaches infinity at infinity, this is a flow case with external boundaries. It is the case of fluid flow within a hollow cylinder filled with fluid and rotating with angular velocity ω about its axis. The flow net is shown in Fig. 72.

Fig. 73.—Flow net for rotation of elliptic cylinder about its axis in an infinite fluid.

74. Rotation of an Elliptic Cylinder in an Infinite Fluid. The fluid motion due to the rotation of an elliptic cylinder about its axis in an infinite fluid is given by the complex potential

$$w = iCe^{-2\zeta}, \qquad z = c \cosh \zeta \qquad (67)$$

The potential and stream functions are

$$\phi = Ce^{-2\xi} \sin 2\eta, \qquad \psi = Ce^{-2\xi} \cos 2\eta$$

Expressing the boundary condition [Eq. (65)] in elliptic coordinates,

$$\psi = \tfrac{1}{4}c^2\omega\,(\cosh 2\xi + \cos 2\eta) + D$$

where D is the arbitrary constant. Substituting this into the stream function,

$$Ce^{-2\xi} \cos 2\eta = \tfrac{1}{4}c^2\omega(\cosh 2\xi + \cos 2\eta) + D$$

This equation is satisfied by $\xi = \xi_0$, provided

$$Ce^{-2\xi_0} = \tfrac{1}{4}c^2\omega, \qquad \tfrac{1}{4}c^2\omega \cosh 2\xi_0 + D = 0$$

Letting a, b be the semimajor and semiminor axes of the elliptic cylinder $\xi = \xi_0$,

$$a = c \cosh \xi_0, \qquad b = c \sinh \xi_0$$

and since

$$e^{\xi_0} = \sinh \xi_0 + \cosh \xi_0 = \frac{a + b}{c}$$

then

$$C = \tfrac{1}{4}\omega(a + b)^2$$

C may be substituted back into Eqs. (67) to give the complex potential. As ξ approaches infinity, the potential function approaches a constant value. As ξ approaches infinity at an infinite distance from the origin, all necessary conditions are fulfilled for rotation of an elliptic cylinder in an infinite fluid. The flow net is given in Fig. 73.

Setting $b = 0$ reduces this case to the rotation of a rectangular lamina in an infinite fluid. The same flow net applies as for the elliptic cylinders.

Exercises

1. Construct a flow net for flow along two planes intersecting at 135-deg angle.

2. Sketch the flow net for two sources of equal strength located at $(1,0)$ and $(-1,0)$. What is the complex potential? Find all the singular points in the finite z-plane. Indicate all planes of symmetry.

3. Show, from the definitions, that Eqs. (1), (5), (10), (13), (20), (25), and (27) are true.

4. Sketch the flow net for the transformation $w = (1 + i)z^2$.

5. Find the complex potential for 10 ft per sec approach velocity flowing through a grating of parallel cylindrical bars 2 in. in diameter spaced 1 ft center to center. Are there any singular points in the flow pattern outside the cylinders?

6. Select the functional relation between z and w to portray a flow of 100 ft³ per sec out of a channel with parallel walls 10 ft apart. Consider a depth of 1 ft. Sketch the flow net. *Ans.* $10z = -w + Ce^{-\pi w/50}$.

7. What is the potential function for an elliptic cylinder translating parallel to its major axis?

8. Find the kinetic energy of the fluid in Exercise 7.

9. Find the virtual mass for an elliptic cylinder translating parallel to its major axis.

10. Work out an expression for the fluid velocity at the surface of an elliptic cylinder translating parallel to its major axis.

11. Investigate the function

$$w = \frac{i\kappa\zeta}{2\pi}, \qquad z = c \cosh \zeta$$

What is the flow pattern obtained by superposing this upon steady flow around an elliptic cylinder?

CHAPTER VII

BLASIUS THEOREM—FLOW AROUND CYLINDERS AND AIRFOILS

The theory and use of complex variables have been examined in Chaps. V and VI. In this chapter examples requiring multiple transformations are worked out, leading to irrotational flow around an airfoil with circulation.

75. Resultant Fluid Forces and Moments on Cylinders. The Blasius Theorem. When the complex potential for flow around any cylinder is known, the resultant fluid forces and moments may be determined by use of the Blasius theorem. Referring to Fig. 74, the pressure force acting on an element of the surface is $p\,ds$ and is normal to the surface element ds. Unit width is assumed. The components of the fluid pressure force dX, dY in the positive x- and y-directions may be expressed as follows:

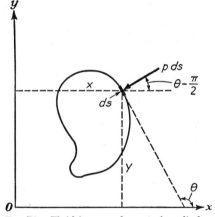

FIG. 74.—Fluid force on element of a cylinder.

$$dX = -p\,ds\,\cos\left(\theta - \frac{\pi}{2}\right)$$

$$dY = -p\,ds\,\sin\left(\theta - \frac{\pi}{2}\right)$$

where θ is the angle the element makes with the x-axis. The differential fluid force on the element may be expressed as a complex number:

$$dF = dX + i\,dY = -p\,ds\left[\cos\left(\theta - \frac{\pi}{2}\right) + i\sin\left(\theta - \frac{\pi}{2}\right)\right]$$

$$= -p\,ds\,e^{i(\theta - \pi/2)}$$

or

$$dF = ip\,ds\,e^{i\theta}$$

using DeMoivre's theorem (Sec. 46). Since $dz = ds\,e^{i\theta}$, the conjugate of dF is

$$d\bar{F} = dX - i\,dY = -ip\,ds\,e^{-i\theta} \qquad (1)$$

$$= -ip\,dz\,e^{-i2\theta}$$

137

The moment about the origin due to fluid pressure on the element is

$$-y\,dX + x\,dY = p\,ds\left[y\cos\left(\theta - \frac{\pi}{2}\right) - x\sin\left(\theta - \frac{\pi}{2}\right)\right]$$
$$= p\,ds(y\sin\theta + x\cos\theta)$$

which is the real part of

$$pz\,dz\,e^{-i2\theta} = iz\,d\bar{F}$$

Therefore, the moment about the origin dN due to fluid pressure is given by

$$dN + i\,dM = iz\,d\bar{F} \tag{2}$$

where dM is the pure imaginary part of $iz\,d\bar{F}$.

Integrating Eqs. (1) and (2) around the closed cylinder

$$\bar{F} = X - iY = -i\oint pe^{-i2\theta}\,dz \tag{3}$$

and

$$N + iM = \oint pe^{-i2\theta}z\,dz \tag{4}$$

where the small circle in the integral sign indicates that the integration is to be carried out completely around the periphery of the cylinder.

The Blasius theorem is derived for steady flow; hence, Bernoulli's equation may be written

$$p = c - \frac{\rho}{2}q^2 \tag{5}$$

with extraneous forces omitted. The constant c cannot affect either the resultant force or moment on the cylinder. It can be dropped out; thus,

$$p = -\frac{\rho}{2}q^2 \tag{6}$$

The complex velocity is

$$-\frac{dw}{dz} = u - iv$$

Since the fluid velocity must be tangent to the cylinder,

$$u - iv = qe^{-i\theta}$$

and

$$q = (u - iv)e^{i\theta} = -\frac{dw}{dz}e^{i\theta} \tag{7}$$

Substituting Eq. (7) in Eq. (6) and Eq. (6) in Eqs. (3) and (4), the Blasius theorem is obtained in equation form:

$$X - iY = \frac{i}{2} \rho \oint \left(\frac{dw}{dz}\right)^2 dz \tag{8}$$

and

$$N + iM = -\frac{\rho}{2} \oint z \left(\frac{dw}{dz}\right)^2 dz \tag{9}$$

These expressions are most easily evaluated by use of the Cauchy integral theorem, which is proved in the following section.

76. The Cauchy Integral Theorem.   The Cauchy integral theorem is the most fundamental theorem in function theory. Let $f(z)$ be a regular function, as defined in Sec. 49, in a simply connected region S, and let C be a closed curve in this region.   The Cauchy integral theorem states that the line integral of $f(z)$ around any closed curve in this region is zero, or

$$\oint_C f(z) \, dz = 0 \tag{10}$$

To prove the theorem, let $f(z) = \xi + i\eta$ and $z = x + iy$, where ξ and η are the real and pure imaginary parts of $f(z)$.   Then

$$f(z) \, dz = (\xi + i\eta)(dx + i \, dy)$$

and Eq. (10) becomes

$$\oint (\xi \, dx - \eta \, dy) + i \oint (\eta \, dx + \xi \, dy)$$

Referring back to Stokes' theorem [Eq. (29), Sec. 24] and letting $R = 0$, $P = \xi$, $Q = -\eta$ yields

$$\oint (\xi \, dx - \eta \, dy) = - \int \int \left(\frac{\partial \xi}{\partial y} + \frac{\partial \eta}{\partial x}\right) dx \, dy \tag{11}$$

Then letting $R = 0$, $P = \eta$, $Q = \xi$,

$$\oint (\eta \, dx + \xi \, dy) = - \int \int \left(\frac{\partial \eta}{\partial y} - \frac{\partial \xi}{\partial x}\right) dx \, dy \tag{12}$$

But ξ and η satisfy the Cauchy-Riemann equations [Eqs. (5), Sec. 49]. Hence, the right-hand side of Eqs. (11) and (12) are zero and the theorem [Eq. (10)] is proved.

When one or more singular points are present in the region enclosed by the curve, the integral $\oint f(z) \, dz$ may not vanish.   An important extension of the Cauchy integral theorem is as follows: Let $f(z)$ be a regular function in a multiply connected region[1] S, and let C_1 and C_2 be closed curves in this region that can be continuously deformed into each other without leaving the region (Fig. 75). Then

$$\oint_{C_1} f(z)\,dz = \oint_{C_2} f(z)\,dz \tag{13}$$

where the line integrals are taken in the same direction, either clockwise, or counterclockwise.

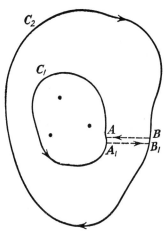

PROOF: Consider the path of integration, say C_3, shown in Fig. 75 by the arrows. C_3 is simply connected; hence, integrating around it, beginning at A,

$$\oint_{C_1} f(z)\,dz + \int_{A_1}^{B_1} f(z)\,dz$$
$$+ \oint_{C_2} f(z)\,dz + \int_{B}^{A} f(z)\,dz = 0$$

by the Cauchy integral theorem. Letting A approach A_1 and B approach B_1, the second and fourth integrals neutralize each other, whence Eq. (13) is proved.

77. Evaluation of the Blasius Theorem. Using the Cauchy integral theorem and its extension, Eqs. (8) and (9) can be further reduced.

FIG. 75.—Extension of a curve in the regular region.

Lift. When the square of the derivative of the potential function $w = f(z)$ is expressed by the series

$$\left(\frac{dw}{dz}\right)^2 = A_0 + \frac{A_1}{z} + \frac{A_2}{z^2} + \cdots \tag{14}$$

for large values of z, Eq. (8) can be integrated by choosing a large circular path of radius R with center at origin.

$$X - iY = i\,\frac{\rho}{2} \oint \left(A_0 + \frac{A_1}{z} + \frac{A_2}{z^2} + \cdots \right) dz$$

Letting $z = Re^{i\theta}$, $dz = iRe^{i\theta}\,d\theta$,

$$X - iY = i\,\frac{\rho}{2} \int_0^{2\pi} \left(iA_0 Re^{i\theta} + iA_1 + i\,\frac{A_2}{R} e^{-i\theta} + \cdots \right) d\theta$$
$$= i\,\frac{\rho}{2} \left(A_0 Re^{i\theta} + iA_1\theta - \frac{A_2}{R} e^{-i\theta} - \cdots \right)_0^{2\pi}$$
$$= -\pi\rho A_1 \tag{15}$$

---

[1] A region is said to be multiply connected when there are closed curves in the region that cannot be reduced to a point without leaving the region. This is the case when part of the boundary of the region is interior to the region, as in Fig. 75.

where A_1, usually complex, is determined from the particular complex potential used.

Moment. To find the moment about the origin due to fluid pressure forces, from Eq. (9),

$$N + iM = -\frac{\rho}{2} \oint z \left(A_0 + \frac{A_1}{z} + \frac{A_2}{z^2} + \cdots \right) dz$$

for large values of z. Again letting $z = Re^{i\theta}$,

$$
\begin{aligned}
N + iM &= -\frac{\rho}{2} \int_0^{2\pi} (iA_0 R^2 e^{2i\theta} + iA_1 Re^{i\theta} + iA_2 + \cdots)\, d\theta \\
&= -\frac{\rho}{2} \left(\frac{A_0 R^2}{2} e^{2i\theta} + A_1 Re^{i\theta} + iA_2 \theta + \cdots \right)_0^{2\pi} \\
&= -iA_2 2\pi\rho
\end{aligned}
\tag{16}
$$

Hence, the moment N is the real part of $-iA_2 2\pi\rho$, where A_2 in general is complex.

The evaluation of these equations assumes no singularities in the fluid outside the cylinder. Equations (15) and (16) are evaluated for particular flow cases later in the chapter.

78. Steady Flow around a Circular Cylinder without Circulation. The complex potential $w = Uz$ is for uniform flow with velocity U in the negative x-direction. The complex potential $w = Ua^2/z$ is for a doublet at the origin with axis in the positive x-direction. The superposition of the uniform flow upon the doublet yields steady flow around a circular cylinder. Expressing z in polar coordinates,

$$w = U \left(z + \frac{a^2}{z} \right) = U \left(re^{i\theta} + \frac{a^2}{r} e^{-i\theta} \right)$$

which may be separated into the components ϕ, ψ, thus

$$\phi = U \left(r + \frac{a^2}{r} \right) \cos\theta, \qquad \psi = U \left(r - \frac{a^2}{r} \right) \sin\theta \tag{17}$$

The streamline $\psi = 0$ is given by $\theta = 0$, π, and by $r = a$, i.e., by the x-axis and by the cylinder $r = a$.

The complex velocity is

$$-\frac{dw}{dz} = -U + \frac{a^2 U}{z^2}$$

showing the uniform velocity $u = -U$ at great distances from the cylinder. To find A_1, A_2 in Eq. (14)

$$\left(\frac{dw}{dz} \right)^2 = U^2 - \frac{2a^2 U^2}{z^2} + \frac{a^4 U^2}{z^4}$$

from which $A_1 = 0$, $A_2 = -2a^2U^2$. Hence, from Eqs. (15) and (16) the resultant force and moment on the cylinder is zero. Stagnation points occur at $x = \pm a$, $y = 0$. The flow net is shown in Fig. 76.

79. Steady Flow around a Circular Cylinder with Circulation. The complex potential

$$w = \frac{i\kappa}{2\pi} \ln z = \frac{i\kappa}{2\pi} \ln (re^{i\theta}) \tag{18}$$

is for circulation κ about the origin in the positive (counterclockwise) direction (Sec. 56). Separating into real and pure imaginary portions,

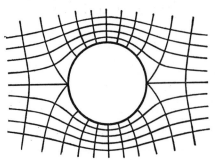

$$\phi = -\frac{\kappa}{2\pi} \theta, \qquad \psi = \frac{\kappa}{2\pi} \ln r$$

from which the streamlines are seen to be circles concentric with the origin and the equipotential lines straight lines through the origin.

Fig. 76.—Flow pattern for uniform flow around a circular cylinder without circulation.

Superposing this flow upon the steady flow around a circular cylinder of the preceding section,

$$\left. \begin{aligned} w &= U\left(z + \frac{a^2}{z}\right) + \frac{i\kappa}{2\pi} \ln z \\ \frac{dw}{dz} &= U\left(1 - \frac{a^2}{z^2}\right) + \frac{i\kappa}{2\pi z} \end{aligned} \right\} \tag{19}$$

The potential and stream functions are

$$\phi = U\left(r + \frac{a^2}{r}\right) \cos \theta - \frac{\kappa}{2\pi} \theta \tag{20}$$

$$\psi = U\left(r - \frac{a^2}{r}\right) \sin \theta + \frac{\kappa}{2\pi} \ln r \tag{21}$$

The streamline $\psi = (\kappa/2\pi) \ln a$ is the circular cylinder $r = a$, showing that this is still the case of uniform flow around a circular cylinder. From the complex velocity [Eqs. (19)], the flow is $u = -U$ at great distances from the cylinder.

The velocity at the surface of the cylinder, necessarily tangent to the cylinder, is

$$q = \frac{\partial \psi}{\partial r_{r=a}} = 2U \sin \theta + \frac{\kappa}{2\pi a}$$

Stagnation points occur where $q = 0$; *i.e.*,

$$\sin \theta = - \frac{\kappa}{4\pi U a}$$

When the circulation is equal to $4\pi U a$, the two stagnation points coincide at $r = a$, $\theta = -\pi/2$. For larger circulation the stagnation points move out into the fluid.

The pressure intensity at the surface of the cylinder is, from Eq. (22), Sec. 10,

$$p = \frac{\rho}{2}(q_0{}^2 - q^2)$$

$$= \frac{\rho}{2} U^2 \left[1 - \left(2 \sin \theta + \frac{\kappa}{2\pi a U} \right)^2 \right]$$

To evaluate the resultant fluid force and moment about the origin, the coefficients A_1 A_2, in Eq. (14) are determined from the complex potential [Eqs. (19)].

$$\left(\frac{dw}{dz} \right)^2 = U^2 + i \frac{U\kappa}{\pi z} - \left(2U^2 a^2 + \frac{\kappa^2}{4\pi^2} \right) \frac{1}{z^2} - i \frac{U a^2 \kappa}{\pi z^3} + \frac{U^2 a^4}{z^4}$$

from which

$$A_1 = \frac{iU\kappa}{\pi}, \qquad A_2 = - \left(2U^2 a^2 + \frac{\kappa^2}{4\pi^2} \right)$$

Hence, the resultant force is

$$X = 0, \qquad Y = \rho U \kappa$$

There is no drag force in the direction of flow, but a force at right angles to the flow equal to the product of fluid density, circulation, and approach velocity. This thrust is referred to as *Magnus effect*. The Flettner rotor ship was designed to utilize this principle by mounting circular cylinders with axes vertical on a ship, then mechanically rotating the cylinders. Air flowing around the rotors produces the thrust at right angles to the relative wind direction. Since A_2 is real, $N = 0$ and no moment is developed.

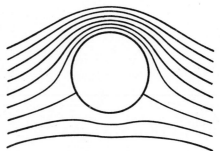

Fig. 77.—Streamlines for uniform flow around a circular cylinder with circulation.

The flow pattern is shown in Fig. 77. It is apparent from the spacing of streamlines that the velocity is lower along the lower portion of the cylinder and therefore the pressure is greater than along the upper portion of the cylinder.

80. Flow around a Circular Arc. A series of transformations due to Kutta and Joukowski are given in this section and developed for the special case of flow around a circular arc with circulation. By a change in one transformation the same relations give equations for flow around a Joukowski airfoil.

The first transformation is from the rectangular $\phi\psi$-grid of the w-plane to flow about a circular cylinder of radius a with uniform approach velocity U in the negative x-direction, and arbitrary circulation κ. This plane is the z''-plane. The complex potential is

$$w = U\left(z'' + \frac{a^2}{z''}\right) + \frac{i\kappa}{2\pi}\ln z'' \tag{22}$$

It may be noted that at great distances from the origin the flow patterns are identical in the w- and z''-planes. The second transformation, from the z''-plane to the z'-plane (Fig. 78), provides for a uniform flow U from an arbitrary direction α. It is

$$z'' = z'e^{i\alpha} \tag{23}$$

where α is the angle the approach velocity makes with the negative x-axis, as shown in Fig. 78. The third transformation provides a transfer of origin. It is to the ζ-plane, given by

$$z' = \zeta - me^{i\delta} \tag{24}$$

where m is the distance $\overline{OO'}$ and δ is equal to $\pi/2$ for the circular arc. The final transformation is an inverse one:

$$z = \zeta + \frac{b^2}{\zeta} \tag{25}$$

By proper selection of the constants in the transformations, flow about any of the Joukowski airfoils can be obtained.

Letting $\delta = \pi/2$ and selecting m and b so that the points A, B in the z'-plane transform into the points A, B on the real axis at $(\pm b, 0)$ in the ζ-plane, the circle is transformed into a circular arc. From Fig. 78,

$$m = a\cos\beta, \qquad b = a\sin\beta$$

To prove that the circle is transformed into a circular arc, Eq. (25) may be written in the following two forms:

$$(z - 2b)\zeta = (\zeta - b)^2, \qquad (z + 2b)\zeta = (\zeta + b)^2$$

Dividing the first by the second,

$$\frac{z - 2b}{z + 2b} = \frac{(\zeta - b)^2}{(\zeta + b)^2}$$

Expressing the four complex numbers making up this ratio in exponential

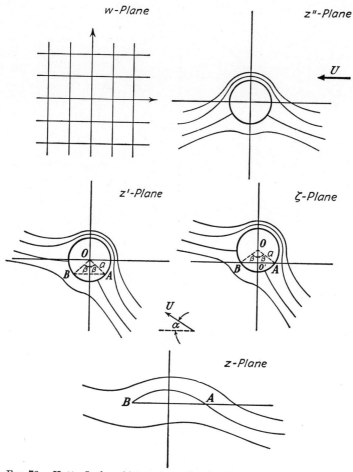

Fig. 78.—Kutta-Joukowski transformation for flow around a circular arc.

form,

$$\zeta - b = r_1 e^{i\theta_1}, \qquad \zeta + b = r_2 e^{i\theta_2}$$
$$z - 2b = r_1' e^{i\theta_1'}, \qquad z + 2b = r_2' e^{i\theta_2'}$$

the ratio becomes

$$\frac{r_1'}{r_2'} e^{i(\theta_1' - \theta_2')} = \frac{r_1^2}{r_2^2} e^{i2(\theta_1 - \theta_2)}$$

The arguments of these two complex numbers must be equal; hence,

$$\theta_1' - \theta_2' = 2(\theta_1 - \theta_2) \qquad (26)$$

Let P be a point on the circle in the ζ-plane (Fig. 79). Then as P moves in a counterclockwise direction around the circle starting at A,

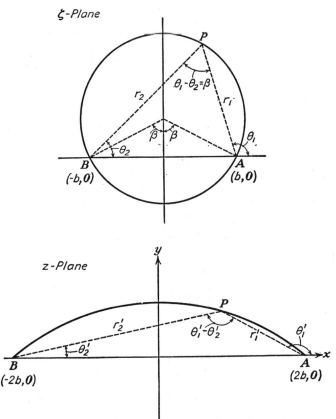

Fig. 79.—Transformation of circle into a circular arc.

the corresponding point P in the z-plane describes a circular arc, as $\theta_1' - \theta_2'$ is constant and equal to $2(\theta_1 - \theta_2)$ by Eq. (26). From the geometry of the circle $\beta = \theta_1 - \theta_2$. As P passes B in the ζ-plane, $\theta_1 - \theta_2$ increases by π and hence $\theta_1' - \theta_2'$ increases by 2π. Therefore, as P completes the circle in the ζ-plane, it moves back along the arc in the z-plane.

The complex velocity for flow around the circular arc is given by

$$\frac{dw}{dz} = \frac{dw}{dz''}\frac{dz''}{dz'}\frac{dz'}{d\zeta}\frac{d\zeta}{dz} = \frac{dw}{dz''}\frac{e^{i\alpha}}{1 - b^2/\zeta^2}$$

$$= -u + iv$$

The points $\zeta = \pm b$ are singular points having infinite velocity unless $\dfrac{dw}{dz''}$ equals zero at these points. These are the points A,B. By taking one of them as a stagnation point in the z''-plane, the flow is made finite there, but in general it is impossible to make the flow finite at both A,B at the same time. To make A the stagnation point, from Eq. (22),

$$\frac{dw}{dz''} = U\left(1 - \frac{a^2}{z''^2}\right) + \frac{i\kappa}{2\pi z''} = 0 \qquad (27)$$

for the point A, which can be expressed

$$z' = z''e^{-i\alpha} = a\sin\beta - ia\cos\beta = -iae^{i\beta}$$

Then

$$z'' = -iae^{i(\beta+\alpha)}$$

Substituting into Eq. (27) and solving for κ,

$$\kappa = 4\pi a U \cos(\alpha + \beta) \qquad (28a)$$

In a similar manner, if B is the stagnation point,

$$\kappa = 4\pi a U \cos(\alpha - \beta) \qquad (28b)$$

The velocity is necessarily tangent to the arc when the circulation is selected to make it finite at the end.

From Fig. 80,

$$R = a\sec\beta$$

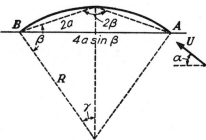

By selecting the desired chord $4a$ $\sin\beta$ and radius R, the values of a and β may be determined. The approach velocity U and its direction are arbitrary. The circulation is selected from Eqs. (28) for finite velocity at A.

Fig. 80.—Circular arc showing values of constants in transformations.

To find the resultant fluid force and moment about the origin, the constants A_1, A_2 in Eq. (14) are required. Using Eqs. (22) to (25),

$$\frac{dw}{dz''} = U\left(1 - \frac{a^2}{z''^2}\right) + \frac{i\kappa}{2\pi z''} = U\left[1 - \frac{a^2 e^{-i2\alpha}}{(\zeta - me^{i\delta})^2}\right] + \frac{i\kappa e^{-i\alpha}}{2\pi(\zeta - me^{i\delta})}$$

since

$$z'' = (\zeta - me^{i\delta})e^{i\alpha}$$

Expanding $\dfrac{dw}{dz''}$ in series of descending powers of ζ,

$$\frac{dw}{dz''} = U + \frac{i}{\zeta}\frac{\kappa e^{-i\alpha}}{2\pi} + \frac{1}{\zeta^2}\left(\frac{i\kappa m}{2\pi}e^{i(\delta-\alpha)} - a^2 U e^{-i2\alpha}\right)$$

$$+ \frac{1}{\zeta^3}\left(\frac{i\kappa m^2}{2\pi}e^{i(2\delta-\alpha)} - 2a^2 U m e^{i(\delta-2\alpha)}\right) + \cdots$$

Also

$$\frac{d\zeta}{dz} = \left(1 - \frac{b^2}{\zeta^2}\right)^{-1} = 1 + \frac{b^2}{\zeta^2} + \frac{b^4}{\zeta^4} + \cdots$$

and

$$\frac{dz''}{dz'} = e^{i\alpha}, \qquad \frac{dz'}{d\zeta} = 1$$

Combining,

$$\frac{dw}{dz} = \frac{dw}{dz''}\frac{dz''}{dz'}\frac{dz'}{d\zeta}\frac{d\zeta}{dz}$$

and expressing in series form,

$$\frac{dw}{dz} = U e^{i\alpha} + \frac{i\kappa}{2\pi\zeta} + \frac{1}{\zeta^2}\left(U e^{i\alpha}b^2 + \frac{i\kappa m}{2\pi}e^{i\delta} - a^2 U e^{-i\alpha}\right) + \cdots$$

From Eq. (25)

$$\zeta = z - \frac{b^2}{z} - \frac{b^4}{z^3} - \cdots$$

hence,

$$\frac{1}{\zeta} = \frac{1}{z} + \frac{b^2}{z^3} + \cdots$$

and

$$\frac{1}{\zeta^2} = \frac{1}{z^2} + \frac{2b^2}{z^4} + \cdots$$

Substituting these in the expression for $\dfrac{dw}{dz}$,

$$\frac{dw}{dz} = U e^{i\alpha} + \frac{i\kappa}{2\pi z} + \frac{1}{z^2}\left(U e^{i\alpha}b^2 + \frac{i\kappa m}{2\pi}e^{i\delta} - a^2 U e^{-i\alpha}\right) + \cdots$$

Squaring,

$$\left(\frac{dw}{dz}\right)^2 = U^2 e^{2i\alpha} + \frac{i\kappa U e^{i\alpha}}{\pi z} + \frac{1}{z^2}\left(-\frac{\kappa^2}{4\pi^2} + 2U^2 e^{i2\alpha}b^2 + \frac{iU\kappa m}{\pi}e^{i(\alpha+\delta)} - 2U^2 a^2\right)$$

$$+ \cdots$$

Therefore,

$$A_1 = \frac{i\kappa U e^{i\alpha}}{\pi}$$

and

$$A_2 = 2U^2 e^{i2\alpha} b^2 - \frac{\kappa^2}{4\pi^2} + i\frac{U\kappa m}{\pi} e^{i(\alpha+\delta)} - 2U^2 a^2$$

From Eq. (15) the fluid force components exerted on the body resulting from this series of transformations are

$$X - iY = -\pi\rho \left(\frac{i\kappa U e^{i\alpha}}{\pi} \right) = \rho\kappa U \sin \alpha - i\rho\kappa U \cos \alpha$$

or

$$X = \rho\kappa U \sin \alpha, \qquad Y = \rho\kappa U \cos \alpha \qquad (29)$$

These are the components of a force at right angles to the undisturbed stream U. Hence, the body is subjected to a lift force

$$L = \rho\kappa U \qquad (30)$$

The moment about the origin, from Eq. (16), experienced by the body resulting from this transformation, is

$$M_0 = 2\pi\rho U^2 b^2 \sin 2\alpha + \rho U\kappa m \cos (\alpha + \delta) \qquad (31)$$

Referring back to the circular arc, a special case of this transformation, the lift for finite velocity at A is, from Eqs. (28) and (30),

$$L = 4\pi\rho a U^2 \cos (\alpha + \beta)$$

and the moment, since $\delta = \pi/2$, is

$$M_0 = 2\pi\rho U^2 b^2 \sin 2\alpha - 4\pi\rho a U^2 m \sin \alpha \cos (\alpha + \beta)$$

If it is desired that the finite velocity be at the trailing edge, the value of κ is determined from Eq. (28b).

Example: What is the lift per foot of span on a thin lamina bent in the shape of a circular arc of radius 6 ft and length along the arc of 4 ft? The velocity of lamina is 10 ft per sec, and the fluid is salt water. $\rho = 1.99$ slugs per ft³.

Solution: From Fig. 80, $\gamma = \frac{2}{6} = 0.333$ radians and $2\beta + \gamma = \pi$; hence,

$$\beta = \pi/2 - \tfrac{1}{6} = 1.404 \text{ radians}$$
$$a = R \cos \beta = 6 \cos (1.404) = 0.99 \text{ ft.}$$

$\kappa = 4\pi \cdot 0.99 \cdot 10 \cos (1.404 + \alpha) = 124.5 \cos (1.404 + \alpha)$ from Eq. (28) for finite velocity at A. The lift force is

$$L = \rho U\kappa = 2480 \cos (1.404 + \alpha) \text{ lb per ft span}$$

At zero angle of attack ($\alpha = 0$), the lift is 409 lb. If the velocity is made finite at the trailing edge, for zero angle of attack ($\alpha = \pi$) the lift force remains 409 lb.

The angle of zero lift for finite velocity at the trailing edge is given by

$$\cos (1.404 - \alpha) = 0$$

from which $\alpha = -9°40'$ or $170°20'$. The lift equation does not hold in an actual fluid for large values of α. The streamline in contact with the arc separates from the arc, and the lift force drops.

The moment about the origin is

$$M_0 = 1250(0.978 \sin 2\alpha - 0.318 \sin^2 \alpha)$$
$$= 1222(\sin 2\alpha - 0.325 \sin^2 \alpha)$$

81. Joukowski Airfoil. In the preceding section it is shown that the circle of radius a transformed into a circular arc. Now, by selecting the

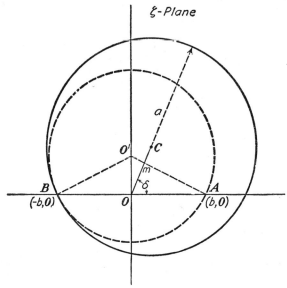

Fig. 81.—Selection of axes in the ζ-plane for the Joukowski airfoil.

proper values of b, δ, and m, the transformation is from a circle into a Joukowski airfoil. The two singular points, $\zeta = \pm b$, are on the real axis and are equidistant from the imaginary axis. In Fig. 81 the dashed circle would transform into a circular arc. The circle of radius a, however, now transforms into a cylinder enclosing A and having a cusp at B, since B is a singular point. The amount of camber of the arc and the cylinder depends upon the ratio $\overline{OO'}/a$, while the thickness of the cylindrical section depends upon δ. a determines the size of the cylinder. That the circle does transform into an airfoil section can be verified by substituting points into Eqs. (22) to (25). A graphical method for constructing the airfoils from known values of m, δ, and a is more expedient. It is explained in the following paragraphs.

The transformation

$$z = \zeta + \frac{b^2}{\zeta}$$

may be regarded as two separate transformations:

$$\zeta_1 = \frac{b^2}{\zeta}, \qquad z = \zeta + \zeta_1$$

With ζ and ζ_1 known the second transformation reduces to simple vector addition.

The first transformation $\zeta_1 = b^2/\zeta$ can be accomplished as follows: For some point, say P, the complex number $\overline{OP} = re^{i\theta} = \zeta$ (Fig. 82). Then

$$\zeta_1 = \frac{b^2}{\zeta} = \frac{b^2}{re^{i\theta}} = \frac{b^2}{r} e^{-i\theta}$$

which is another complex number with modulus b^2/r, the inverse of P in the circle of radius $b = \overline{OA}$, and with amplitude $-\theta$. $\overline{OP'}$ is the

complex number $(b^2/r)e^{i\theta}$, and $\overline{OP_1}$ is the complex number $(b^2/r)e^{-i\theta}$; $\overline{OP_1}$ is said to be the image of $\overline{OP'}$ in the real axis. The graphical procedure is then to take the inverse of P in the circle of radius b, then its image in the real axis. The second transformation $z = \zeta + \zeta_1$ is the vector addition of $\overline{OP}$ and $\overline{OP_1}$, giving $\overline{OR}$ in the z-plane. This is a tedious process when many points are to be plotted.

If P describes a circle, then its inverse in the circle of radius b also describes a circle. Since the

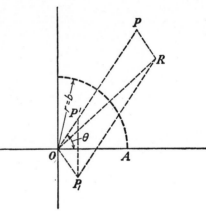

FIG. 82.—Graphical transformation
$$z = \zeta + \frac{b^2}{\zeta}.$$

image of a circle in an axis is another circle, the locus of P_1 is a circle when P describes a circle. To prove this, let P be any point on the circle (center at C) that is to be transformed into an airfoil (Fig. 83). P' is the inverse of P with respect to the circle with center at O and radius b; *i.e.*,

$$\overline{OP'} = \frac{b^2}{\overline{OP}} \tag{32}$$

Extend line PO to cut the circle at Q; draw CQ; then draw a parallel to

CQ through P' to its intersection with CO extended at C'. It is first proved that the locus of P' is a circle whose center is C'.

Since $A'OB$ and POQ are chords of a circle intersecting at O

$$\overline{OP} \times \overline{OQ} = \overline{OA'} \times \overline{OB} = \overline{K^2} \qquad (33)$$

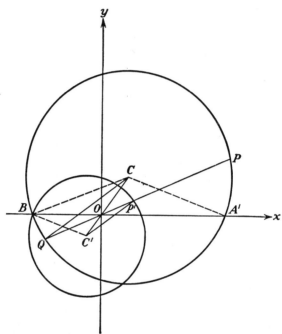

FIG. 83.—Graphical construction for locus of inverse of P.

then from Eqs. (32) and (33),

$$\frac{\overline{OP'}}{\overline{OQ}} = \frac{b^2}{K^2} = \text{constant}$$

The triangles $OP'C'$, OQC are similar, since $P'C'$ was drawn parallel to CQ. Therefore,

$$\frac{\overline{OC'}}{\overline{OC}} = \frac{\overline{C'P'}}{\overline{CQ}} = \frac{\overline{OP'}}{\overline{OQ}} = \frac{b^2}{K^2} = \text{constant}$$

Then, as $\overline{OC}$ is a fixed distance, $\overline{OC'}$ is also a fixed distance and C' is a fixed point. Since $\overline{CQ}$, the radius of the circle, is a fixed length, $\overline{C'P'}$ is also a fixed length. Therefore, P' describes a circle whose center is at C'.

As $\overline{OB} = b$, the point B is its own inverse and the locus of P' passes through B.

Taking P at point B, line BC', parallel to CA', also determines C'. It follows that, as CA' and BC'' are parallel,

$$\angle CA'B = \angle C''BA' = \angle CBA'$$

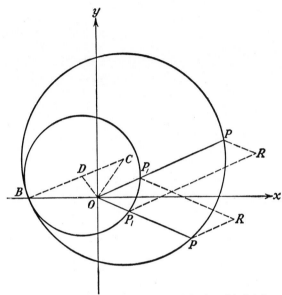

Fig. 84.—Graphical construction of Joukowski airfoil.

Hence, $C'B$ and CB are equally inclined to the real axis. The image of C' in the real axis is on the line CB at point D (Fig. 84) such that

$$\angle DOy = \angle yOC$$

Point B is a point on the reflected circle; therefore, $\overline{DB}$ is the radius.

With the two circles drawn, as in Fig. 84, points on the airfoil section are rapidly constructed by drawing pairs of straight lines through O making equal angles with the real axis. Then corresponding points P, P_1 are added as indicated in the figure.

Symmetrical sections, called *strut sections*, are obtained when C is on the real axis.

82. The Joukowski Hypothesis.   The Joukowski hypothesis is that the circulation for an actual airfoil will always adjust itself so that the velocity is finite at the trailing edge. The hypothesis is fairly well substantiated experimentally. The equations for lift and moment developed in Sec. 80 [Eqs. (28*b*), (30), and (31)] are valid for Joukowski airfoils. For more complete information on the theoretical treatment of airfoil sections, reference is made to the excellent treatment in Glauert, "The Elements of Aerofoil and Airscrew Theory," Macmillan & Co., Ltd., London, 1926.

83. Extended Joukowski Airfoils. Again rewriting Eq. (25) in the form

$$\frac{z - 2b}{z + 2b} = \left(\frac{\zeta - b}{\zeta + b}\right)^2 \tag{34}$$

which transforms the circle into a circular arc or an airfoil, a singular point occurs at $z = 2b$ and $z = -2b$, the end points of the arc. For the Joukowski airfoil the singular point $z = 2b$ is within the contour and does not need any further consideration. At $z = -2b$ a cusp is formed. As it is physically impossible to construct an airfoil with a cusp, a transformation that produces a finite angle at the trailing edge is useful.

In the immediate vicinity of $z = -2b$ let

$$z = -2b + re^{i\theta}, \qquad \zeta = -b + r_1 e^{i\theta_1}$$

where r, r_1 are infinitesimals. Then substituting into Eq. (34),

$$-\frac{4b}{re^{i\theta}} = \frac{4b^2}{r_1^2 e^{i2\theta_1}}$$

Rewriting,

$$\frac{4b}{r} e^{-i(\theta+\pi)} = \frac{4b^2}{r_1^2} e^{-i2\theta_1}$$

and equating arguments of the complex numbers,

$$\theta + \pi = 2\theta_1$$

Hence, as the point traverses the circle in the ζ-plane and passes through B, θ_1 increases by π and θ increases by 2π, resulting in the cusp. By generalizing Eq. (34),

$$\frac{z - nb}{z + nb} = \left(\frac{\zeta - b}{\zeta + b}\right)^n \tag{35}$$

and making similar substitutions for the vicinity of $z = -nb$, $\zeta = -b$, i.e., $z = -nb + re^{i\theta}$, $\zeta = -b + r_1 e^{i\theta_1}$,

$$\theta + (n + 1)\pi = n\theta_1$$

Therefore, by taking $n = 2 - \lambda/\pi$, then increasing θ_1 by π, in moving through B, moves θ through $2\pi - \lambda$, which gives two definite branches at the trailing edge in place of a cusp.   Figure 85 shows an airfoil with a finite angle λ at the trailing edge.   These airfoils are known as extended Joukowski airfoils.

Fig. 85.—Extended Joukowski airfoil.

More general sections may be obtained by generalizing Eq. (25) to the form

$$z = \zeta + \frac{a_1}{\zeta} + \frac{a_2}{\zeta^2} + \cdots$$

where $a_1, a_2, \ldots$ are constants.

Exercises

1. Show that the resultant force and moment of fluid forces on a circular cylinder in a uniform stream must be zero by integration of surface pressure forces around the cylinder.

2. Prove that the resultant force on a circular cylinder in a uniform stream U with circulation κ is $\rho U \kappa$ by taking a surface integral of the pressure force around the cylinder.

3. Find the velocity at the leading edge of a circular arc when the circulation is selected such that the velocity is finite at that point.

Ans. $u = +U \sin \beta \cos 2\beta \sin (\alpha + \beta); v = 2U \cos \beta \sin^2 \beta \sin (\alpha + \beta)$.

4. Graphically construct the Joukowski airfoil determined by $a = 2$ in., $\delta = 60$ deg, $m = 0.4$ in.

5. Find the angle of zero lift for the airfoil section of Exercise 4.   What is the lift when in an air stream ($\rho = 0.00238$ slug per ft³) of velocity 150 ft per sec and $\alpha = 5$ deg?

6. The complex potential at a great distance from a cylinder in a uniform flow may be written

$$w = A + Bz + C \ln z$$

(a) By substitution in Eq. (8) find the resultant force components $X - iY$ in terms of B and C.

(b) Evaluate B and C in terms of U, V, and κ; then show that the resultant force is always a lift force having components

$$X = \kappa \rho V, \qquad Y = -\kappa \rho U$$

CHAPTER VIII

SCHWARZ-CHRISTOFFEL THEOREM   FREE STREAMLINES

The Schwarz-Christoffel theorem provides a method for transforming the flow about a polygon into the uniform flow parallel to the real axis. The assumption of free streamlines permits separation of the flow to take place at those sudden changes in direction of boundaries which cause infinite velocities. As the Schwarz-Christoffel theorem is utilized in studying free streamline transformations, it is derived and examples worked out for special cases in the first part of this chapter.

84. Definitions and Conventions. In hydrodynamical applications leading to free streamlines it is necessary to determine the flow pattern around straight-sided closed figures. Frequently these polygons have vertices at infinity. Rectangles, for example, having two vertices at infinity are referred to as semi-infinite strips, while those with all four vertices at infinity are referred to as infinite strips.

A simple closed polygon is defined as a closed figure composed of straight-line segments such that (1) the boundary may be completely traversed without leaving it, *i.e.*, the boundary is connected, and (2) the boundary divides the whole plane into two regions, one region that is interior to the polygon and the other region that is exterior to the polygon. The interior of the polygon is "connected"; *i.e.*, a path from any point in the interior to any other point in the interior may be followed without crossing a boundary, and similarly for the exterior region. Furthermore, the interior is defined as that region which is on the observer's left as he traverses the boundary in a prescribed sense. For a polygon having all vertices at finite points, the boundary is traversed in the counterclockwise direction to preserve the usual conception of interior. Figure 86 shows several simple closed polygons having vertices at infinity. The subscript ∞ refers to a point that is infinitely distant. The exterior is indicated by hatching. By application of the Schwarz-Christoffel method twice, any simple closed polygon can be transformed into any other simple closed polygon.

85. The Schwarz-Christoffel Theorem. To prove the Schwarz-Christoffel theorem, which states that the interior of a simple closed polygon may be mapped into the upper half of a plane and the boundary

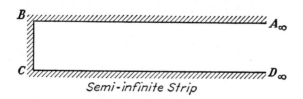

Semi-infinite Strip

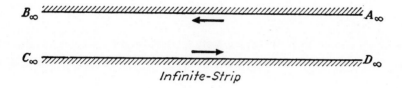

Infinite-Strip

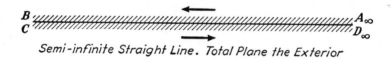

Semi-infinite Straight Line. Total Plane the Exterior

Interior is Upper Half Plane

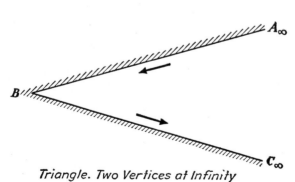

Triangle. Two Vertices at Infinity

Fig. 86.—Examples of simple closed polygons.

of the polygon into the real axis, it is shown that the transformation given by

$$\frac{dz}{dt} = A(a - t)^{-\alpha/\pi}(b - t)^{-\beta/\pi}(c - t)^{-\gamma/\pi} \cdots \tag{1}$$

transforms the real axis in the t-plane into the boundary of a simple closed polygon in the z-plane, where A is a complex constant; $a, b, c, \ldots$ are real constants in ascending order of magnitude; and $\alpha, \beta, \gamma, \ldots$ are angles (positive or negative) such that

$$\alpha + \beta + \gamma + \cdots = 2\pi$$

Plotting $a, b, c, \ldots$ in the t-plane (Fig. 87) and considering Eq. (1), it is clear that $\frac{dz}{dt}$ is not defined at the points $t = a, t = b, t = c, \ldots$ These points are to be excluded from the boundary by small semicircles

FIG. 87.—Real axis of t-plane.

about each singular point, situated on the upper side of the t-axis, as in Fig. 87.

The proof must show (1) that the real axis of t between any two consecutive singular points $a, b, c, \ldots$ transforms into a straight line in the z-plane, (2) that the small semicircles at $a, b, c, \ldots$ transform into small circular arcs subtending the angles $\pi - \alpha, \pi - \beta, \pi - \gamma, \ldots$, respectively, and (3) that the polygon actually closes for large plus and minus values of t.

To prove the first part, consider $\frac{dz}{dt}$ as an operator (Sec. 51) that transforms an element δt in the t-plane into its corresponding element δz in the z-plane. The right-hand side of Eq. (1) is a complex variable, having a fixed modulus and argument for a fixed point in the t-plane. Let the modulus be r and the argument θ; then

$$\frac{dz}{dt} = re^{i\theta}$$

where r, θ vary in general as t varies throughout the t-plane. δz is given by

$$\delta z = \frac{dz}{dt} \delta t = re^{i\theta} \delta t$$

The argument of δz is to be studied as t moves along the real axis. The product of two complex numbers (Sec. 46) is a complex number whose modulus is the product of the moduli and whose argument is equal to the sum of the arguments of the two complex numbers. The argument of δt is always zero, however, as a point moves in the positive direction along the real axis of the t-plane. Hence, the argument of δz must equal the argument of $\dfrac{dz}{dt}$. Examining Eq. (1), $a - t$, $b - t$, $c - t$, . . . are real numbers if a, b, c, . . . are greater than t (t real), and such terms have the constant argument zero. Thus, when t moves in the positive direction along the real axis to the left of a, δz has the argument of A, which is constant, showing that z moves in a straight line. When t is between, say, a and b, $a - t$ is negative and when raised to a fractional power usually becomes complex. In Fig. 88 let

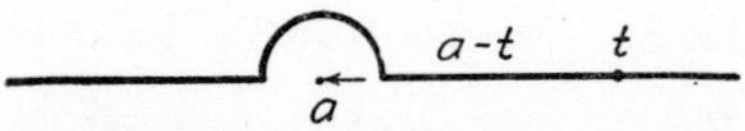

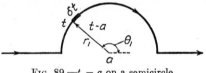

FIG. 88.—$a - t$, for t on the real axis to the right of a.

$$a - t = r_1 e^{i\pi}$$

then

$$(a - t)^{-\alpha/\pi} = r_1^{-\alpha/\pi} e^{-i\alpha}$$

showing its argument to be constant, $-\alpha$. Hence, this portion of the t-axis transforms into a straight line in the z-plane. Similarly, the other portions of the real axis of the t-plane also map into straight lines in the z-plane.

The second part of the proof shows that the small semicircles map into small circular arcs subtending the angles $\pi - \alpha$, $\pi - \beta$, $\pi - \gamma$, ... The transformation of δt into δz will be examined in detail for one of the semicircles, say a. Consider the element δt to be on the arc (Fig. 89). Then

FIG. 89.—$t - a$ on a semicircle.

$$t - a = r_1 e^{i\theta_1}$$

where r_1, the radius, is constant and θ_1, the argument, varies from π to 0. Taking the derivative

$$\delta t = i r_1\, \delta\theta_1 e^{i\theta_1}$$

Writing

$$i = e^{i\pi/2}$$
$$\delta t = r_1\, \delta\theta_1 e^{i[(\pi/2)+\theta_1]}$$

where $r_1 \, \delta\theta_1$ is the modulus and $(\pi/2) + \theta_1$ the argument. δz now becomes

$$\delta z = \frac{dz}{dt} \delta t$$
$$= A(-r_1 e^{i\theta_1})^{-\alpha/\pi}(b - t)^{-\beta/\pi}(c - t)^{-\gamma/\pi} \cdots r_1 e^{i[(\pi/2)+\theta_1]} \, \delta\theta_1$$

Since r_1 can be made as small as desired, $(b - t)^{-\beta/\pi}$, $(c - t)^{-\gamma/\pi}$, . . . become constants in the limit as r_1 approaches zero. Grouping together all the terms that do not contain r_1 or θ_1 into a complex number F,

$$\delta z = F r_1^{1-\alpha/\pi} e^{i\theta_1(1-\alpha/\pi)} \, \delta\theta_1$$

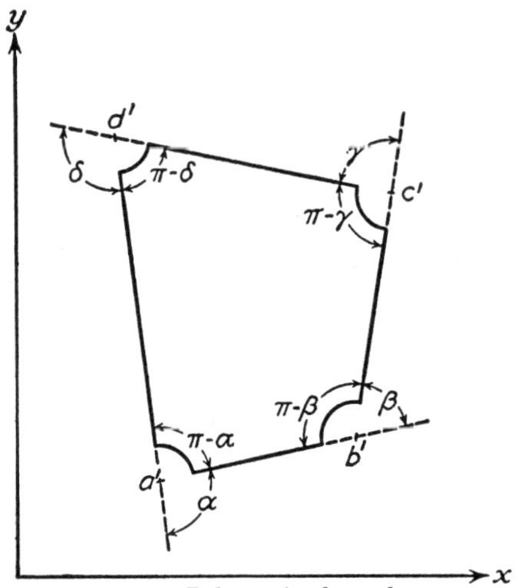

For $|\alpha| < \pi$, which does not restrict the problem since α may be positive or negative, the modulus of δz approaches zero as r_1 approaches zero. When θ_1 changes from π to zero, i.e., through $-\pi$, the argument of δz is decreased by $\pi - \alpha$, which proves the second part. Figure 90 shows the z-plane for a possible polygon.

FIG. 90.—Polygon in the z-plane.

Thus far it has been shown that as a point traverses the real axis of the t-plane, the corresponding point in the z-plane moves in a straight line between the singular points a', b', c', . . . and at each singular point the direction undergoes an angle change (Fig. 90). $\alpha, \beta, \gamma,$ . . . are the exterior angles of the polygon. Since the interior has been kept to the left, the upper half of the t-plane has been transformed into the polygon.

The third part of the proof, to show that the polygon in the z-plane properly closes, requires consideration of the infinite regions of the t-plane (Fig. 91). The upper half of the t-plane is contained within the semicircle of radius R when R is allowed to approach infinity. For t at points along the positive real axis beyond c, assuming a three-sided polygon for convenience, z will follow the straight line from c', which has an argument greater than $b'c'$ by γ. For very large values of t, $a - t$, $b - t$, $c - t$ may each be replaced by $-t$. Then from Eq. (1),

$$\delta z = \frac{dz}{dt} \delta t = A(-t)^{-(\alpha+\beta+\gamma)/\pi} \, \delta t$$

Since $\alpha + \beta + \gamma = 2\pi$, this reduces to

$$\delta z = \frac{A}{t^2}\, \delta t$$

which becomes, upon integration,

$$z = -\frac{A}{t} + E'$$

where E' is the constant of integration and must be the value of z when t approaches infinity along the positive real axis. Hence, it must be a point along the line extending from c'.

In following the motion of z as t traverses the large circle, t may be expressed by

$$t = Re^{i\theta}$$

where R is the modulus of t, also the radius of the circle, and θ is the argument of t. Then, as $a - t$, $b - t$, $c - t$ can still be replaced by $-t$,

$$\delta z = \frac{dz}{dt}\, \delta t = \frac{A}{t^2}\, \delta t = \frac{A}{R^2}\, e^{-i2\theta}\, \delta t$$

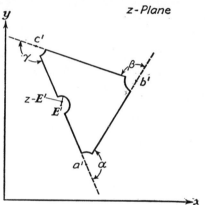

Expressing $A = Ce^{i\lambda}$, where C and λ are constant, and differentiating t for constant R,

$$\delta t = iRe^{i\theta}\, \delta\theta = Re^{i[(\pi/2)+\theta]}\, \delta\theta$$

Making the substitutions in the expression for δz,

$$\delta z = \frac{C}{R}\, e^{i[(\pi/2)+\lambda-\theta]}\, \delta\theta$$

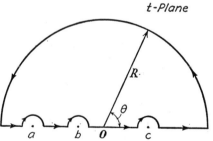

Fig. 91.—Notation for convergence for infinite regions of the t-plane.

Integrating,

$$z = i\frac{C}{R}\, e^{i[(\pi/2)+\lambda-\theta]} + \text{constant}$$

$$= +\frac{C}{R}\, e^{i(\pi+\lambda-\theta)} + \text{constant}$$

To evaluate the constant of integration, as t approaches infinity along the positive t-axis, z approaches E'. Then,

$$R \to \infty \qquad \theta = 0, \qquad z \to E'$$

substitution shows that the constant is E'; thus

$$z = \frac{C}{R} e^{i(\pi+\lambda-\theta)} + E'$$

transforms the large circle in the t-plane onto the z-plane. Rewriting,

$$z - E' = \frac{C}{R} e^{i(\pi+\lambda-\theta)}$$

where C/R is the modulus of $z - E'$. $z - E'$ is a small radius vector with center at E' and argument $\pi + \lambda - \theta$. Therefore, as t describes the infinitely large circle in the t-plane, z describes an infinitesimal arc in the z-plane. The argument of t changes from 0 to π, and the argument of $z - E'$ changes from $\lambda + \pi$ to λ, a change of π in the negative direction.

Now as t leaves the large semicircle at $\theta = \pi$ and traverses the real t-axis from $-\infty$, z moves in a straight line that is an extension of $c'E'$. This line approaches a' as t approaches a; therefore, the polygon in the z-plane closes properly and the side $c'E'a'$ is straight. Letting the small semicircles about the singular points approach zero and the large one approach infinity, the upper half of the t-plane is mapped into a polygon.

This completes the proof of the Schwarz-Christoffel theorem. Integrating Eq. (1),

$$z = A \int \frac{dt}{(a - t)^{\alpha/\pi}(b - t)^{\beta/\pi}(c - t)^{\gamma/\pi} \cdots} + B \qquad (2)$$

where A and B are arbitrary constants. B is a complex constant that determines the location of origin in the z-plane. By proper choice of origin B can be made zero. A is also a complex constant whose modulus affects the scale of the polygon and whose argument determines the orientation of the polygon in the z-plane. Three of the numbers a, b, c, . . . can be selected arbitrarily; the remaining ones are determined by the shape of the polygon.

When the vertex of a polygon corresponds to a point at infinity on the real t-axis, Eq. (1) may be reduced as follows: Let $A = |A|e^{i\lambda}$; then

$$\frac{dz}{dt} = |A|e^{i\lambda}(a - t)^{-\alpha/\pi}(b - t)^{-\beta/\pi}(c - t)^{-\gamma/\pi} \cdots$$

If, for example, $a \to -\infty$, the constant $|A|$ may be written $Ca^{\alpha/\pi}$. Then

$$\frac{dz}{dt} = Ce^{i\lambda}\left(\frac{a - t}{a}\right)^{-\alpha/\pi} (b - t)^{-\beta/\pi}(c - t)^{-\gamma/\pi} \cdots$$

and in the limit as $a \rightarrow -\infty$, the term

$$\left(\frac{a - t}{a}\right)^{-\alpha/\pi}$$

becomes unity and drops out of the equation completely. Hence, for one or more vertices at infinity in the t-plane, the whole factor, including its exterior angle, drops out of the relation.

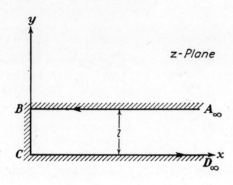

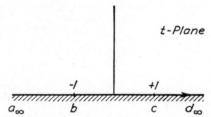

FIG. 92.—Semi-infinite strip mapped into the real axis.

86. Mapping a Semi-infinite Strip into a Half Plane.   As an example of the Schwarz-Christoffel transformation, the semi-infinite rectangle given in the z-plane of Fig. 92 is mapped into the real axis of the t-plane. The exterior angles are

$$\alpha = \beta = \gamma = \delta = \frac{\pi}{2}$$

therefore,

$$z = A \int \frac{dt}{\sqrt{(a - t)(b - t)(c - t)(d - t)}} + B \qquad (3)$$

As three of the points a, b, c, d in the t-plane are arbitrary, they are selected as $t = -\infty$, $t = -1$, $t = +1$, for a, b, and c, respectively.   Evidently, d is also at infinity, since if one vertex of a rectangle is at infinity, at least one more must also be at infinity. The factors in Eq. (3) containing a and d drop out. Substituting for b and c,

$$z = A \int \frac{dt}{\sqrt{t^2 - 1}} + B = A \cosh^{-1} t + B$$

Now this equation would apply equally well to any semi-infinite rectangle in the z-plane. To specialize it for the one in Fig. 92, the constants A and B must be determined. For the origin in the z-plane,

$$z = x + iy = 0, \qquad t = 1$$

then,

$$0 = A \cosh^{-1} 1 + B$$

Since $\cosh^{-1} 1 = 0$, $B = 0$. The value of A depends upon scale and orientation. Letting the width of rectangle be l, then at point B, $z = il$ and $t = -1$, giving

$$il = A \cosh^{-1} (-1)$$

Simplifying,

$$\cosh \frac{il}{A} = -1$$

or

$$\cos \frac{l}{A} = -1$$

from Sec. 57. Hence, $l/A = \pi$, and $A = l/\pi$. The mapping function is

$$z = \frac{l}{\pi} \cosh^{-1} t \qquad (4)$$

The interior of the strip covers the whole t-plane above the real axis.

87. Mapping an Infinite Strip into a Half Plane. An infinite strip (Fig. 93) is a rectangle having four vertices at infinity. Letting B_∞, C_∞ coincide in the t-plane at the origin and selecting A_∞ at $t = -\infty$, then D_∞ must correspond to $t = +\infty$. The transformation becomes

$$z = A \int \frac{dt}{t} + B = A \ln t + B \qquad (5)$$

where a, b, c, d are substituted. Letting the width of strip be l, then at D_∞, $z = \infty + i0$, $t = \infty + i0$. At A_∞, $z = \infty + il$. $t = -\infty + i0$. Substituting in Eq. (5),

$$\infty = A \ln \infty + B$$

and

$$\infty + il = A \ln (-\infty) + B$$

The first equation above is satisfied by $B = 0$.   Since

$$\ln \, (- \infty) = \ln \, (\infty \, e^{i\pi}) = \, \infty \, + \, i\pi$$
$$\infty \, + \, il = A(\infty \, + \, i\pi)$$

from which

$$il = iA\pi, \qquad A = \frac{l}{\pi}$$

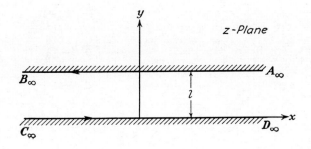

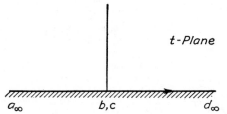

FIG. 93.—Infinite strip mapped into the real axis.

Hence, the mapping function is

$$z = \frac{l}{\pi} \ln t \qquad\qquad (6)$$

for the particular location of axes in Fig. 93.

88. Introductory Remarks concerning Free Streamlines.   When a fluid is made to turn a sudden corner, with radius of curvature zero, the acceleration for the fluid particle becomes infinite.   This calls for an infinite force on the particle, which is obtained in ideal fluid flow by having the velocity go to infinity; *i.e.*, an infinite pressure gradient is set up, with negative infinity pressure at the particle.   Since such a situation has no physical counterpart, the assumption may be made that the fluid particle separates from the boundary rather than make the sharp turn. This assumption leads to the class of problems referred to as *free stream-*

lines. In the remaining sections of this chapter, the assumption is made that separation occurs at those points on the body where the body form makes a sudden turn, with the exception of stagnation points, or those points downstream from the first point of separation.

At separation points in steady flow of a fluid around a body the streamlines leave the body. This dividing streamline is called a *free streamline* in two-dimensional flow, and the fluid in contact with the body downstream from the separation points and separated from the main body of fluid in motion by the free streamlines is known as the *wake.* The fluid in the wake is assumed to be at rest in steady flow problems.

The Blasius theorem (Secs. 75 and 77) does not apply to cases when there is separation. The method of free streamlines provides a drag in irrotational flow of a frictionless fluid around bodies. Due to this and the avoidance of points of infinite velocity, this method permits the solution of many problems that conform closely to similar problems with actual fluids.

The assumption that the fluid is at rest in the wake is considerably in error for actual fluids and frequently leads to a theoretical drag that is much less than in an actual case. When the wake contains another fluid of much less density, the theory should give results comparing favorably with experiment. An example would be the discharge of water out of a slot into air.

In the following sections the effects of gravity are neglected. The pressure intensity in the wake is, therefore, constant, since it is at rest, and the pressure intensity along the free streamline is constant. According to the Bernoulli equation the velocity of the free streamline must also be constant. A streamline in contact with a boundary upstream from the separation point is referred to as a *bounding streamline.* Since the resultant drag on a body is the same whether viewed by an observer as steady or unsteady motion, examples are considered for steady flow only to take advantage of the simpler form of the Bernoulli equation.

When the bounding streamlines are straight, the shape of the free streamlines in two-dimensional motion can be found by the methods of conformal mapping. The transformations are of a special character which takes advantage of the fact that the direction of the bounding streamlines is constant and the speed of the free streamlines is constant. The Schwarz-Christoffel transformations are used and must be familiar to the reader before a complete understanding of the problems can be obtained.

89. Transformations Used in Free-streamline Problems. The transformations required in free-streamline problems are conformal and provide a means of mapping the uniform flow of the w-plane into the free-streamline pattern of the z-plane. Since as many as six complex planes are required for some of the problems, it is a difficult subject. In this section, by means of an example, several of the commonly used transformations are carried out in detail.

It is convenient to assume the velocity of the free streamlines as unity. As long as consistent units are used, it is unimportant what the particular units happen to be. In working with the problem where the free streamline has a 15 ft per sec velocity, the length unit may conveniently be taken as 15 ft and the time unit as 1 sec. All other terms containing lengths should then be expressed accordingly. An example would be pressure intensity, where the consistent unit would be pounds per 225 ft². These remarks apply equally well to other portions of this work.

In free-streamline problems it is convenient to start with the z-plane showing the flow boundaries and the general form of the free streamlines. Then by suitable transformations the bounding streamlines and free streamlines are mapped into straight-sided polygons from which the w-plane is obtained by use of the Schwarz-Christoffel theorem one or more times.

The example selected to illustrate the various steps is the flow out of a tank composed of two flat plates inclined at 45 deg, as shown in the z-plane of Fig. 94. The tank is considered very long normal to the paper, leaving an opening out of which the fluid can flow. The inclined sides are assumed to extend back indefinitely. The bounding streamlines are thus straight; *i.e.*, their argument, or direction, is constant. After the streamline becomes free, it has the special property of constant speed; *viz.*, $u + iv$ has constant modulus.

These special characteristics lead to the transformation from the z-plane to another plane, say the ζ-plane, where

$$\zeta = -\frac{dz}{dw}$$

When ζ is known or can be expressed as a function of z and w, the problem is solved. From Sec. 54,

$$\zeta = -\frac{dz}{dw} = \frac{1}{u - iv} = \frac{1}{q}\left(\frac{u}{q} + i\frac{v}{q}\right) = \left|\frac{1}{q}\right|e^{i\theta} \tag{7}$$

where u, v are component velocities in the x,y directions and q is the resultant velocity in direction θ at a point. $\left|\dfrac{1}{q}\right|$ is the modulus of ζ. Since

$$\left(\frac{u}{q}\right)^2 + \left(\frac{v}{q}\right)^2 = 1$$

it is evident that ζ is a complex number having a modulus which is the reciprocal of speed and an argument which is the angle the velocity vector makes with the positive x-axis.   At B_∞ in the z-plane (Fig. 94) the velocity is zero for $B_\infty A$ very long.   Then ζ has a magnitude infinity

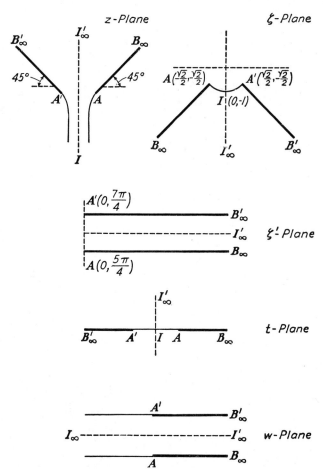

Fig. 94.—Planes used for flow out of a tank with 45-deg sides.

and a direction $\theta = 225$ deg, a point that can be indicated in the ζ-plane. Observing the bounding streamline near A, the argument remains the same, $\theta = 225$ deg, but the modulus becomes unity at A when unit velocity of free streamline is assumed.   The line $B_\infty A$ plots as shown in the ζ-plane.

The free streamline AI changes direction from 225 to 270 deg, but its modulus $1/q$ remains unity. ζ is then the arc of the unit circle from

A to I in the ζ-plane.   Similarly $B'_\infty A'I$ plots in the ζ-plane as in Fig. 94. This transformation has changed the bounding streamlines to other straight lines and has changed the unknown free streamlines into an arc of a circle.   If a means can now be found to change the streamlines in the ζ-plane into a polygon, then by use of the Schwarz-Christoffel method the final conversions can be made to the w-plane.

Considering next the transformation

$$\zeta' = \ln \zeta = \ln \left|\frac{1}{q}\right| + i\theta \tag{8}$$

where ζ' is another complex plane, θ, the argument of the velocity vector is plotted as ordinate against $\ln |1/q|$ as abscissa.   Along the circular arc AIA' in the ζ-plane, $\ln |1/q| = 0$ and θ varies from 225 to 315 deg. Hence, the arc plots as a portion of the imaginary axis.   The line $B_\infty A$ has constant θ; therefore, it plots as a horizontal line starting at A and extending to the right as $\ln |1/q|$ varies from 0 to ∞; likewise for $B'_\infty A'$.

The ζ'-plane has provided a semi-infinite strip that can be mapped into the upper half of another plane, say the t-plane, by the method of Sec. 86.

The t-plane is as shown in Fig. 94.   Since $I'_\infty I$ is also a streamline, another transformation must be made, this time from the t-plane to an infinite strip as worked out in Sec. 87.   This latter transformation has all streamlines parallel to the real axis and is, therefore, the w-plane. The various functions for each transformation being known, ζ may be expressed in terms of z and w only, and the position of the free streamlines can be plotted on the z-plane.

In the following three sections examples are worked out showing various techniques required in working with free streamlines.

90. Borda's Mouthpiece in Two-dimensional Flow.   Borda's mouthpiece, in two dimensions, is a re-entrant slot in a large container, as shown in the z-plane of Fig. 95.   The bounding streamlines AB_∞, $A'B'_\infty$ are assumed to be so long that the velocity at B_∞ and B'_∞ is zero.   The point I'_∞ is in the tank along the line of symmetry and is sufficiently removed from the entrance so that the velocity is also zero; I_∞ is a point on the jet sufficiently far downstream that no more contraction takes place; $I'_\infty I$ is a streamline.

The first transformation is

$$\zeta = \left|\frac{1}{q}\right| e^{i\theta}$$

where θ is the argument of the velocity vector in the z-plane.   Along $B_\infty A$, θ is zero and $|1/q|$ varies from ∞ at B_∞ to unity at A when unit

velocity of free streamlines is assumed.   Then from A the free streamline turns in the negative direction through 180 deg.   Along $B'_\infty A'$ the
argument of the velocity vector is zero also, and $|1/q|$ varies from ∞ at
B'_∞ to unity at A'.   Then the streamline turns through 180 deg in the

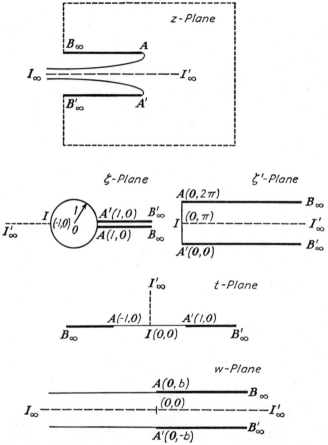

Fig. 95.—Mapping planes for Borda's mouthpiece in two dimensions.

positive direction, with $|1/q|$ unity along the free streamline.   In Fig. 95
the bounding streamlines are indicated by heavy lines.   To avoid con-
fusion, AB_∞ is drawn slightly below $A'B'_\infty$ in the ζ-plane.   At I'_∞, $|1/q|$ is
infinite and $\theta = 180$ deg, plotting as shown in the ζ-plane.

The second transformation

$$\zeta' = \ln \zeta = \ln \left|\frac{1}{q}\right| + i\theta$$

is used to convert the ζ-plane into a semi-infinite strip. A', given by $\zeta = 1$, plots as $\zeta' = 0$, or as origin in the ζ'-plane. Traversing the unit circle, the real part of ζ' remains zero and θ increases to π radians at I and to 2π radians at A, as shown in the ζ'-plane. The point B'_∞ is given by $\zeta = \infty$ or, using the transformation, $\zeta' = \infty$. As ζ is real between A' and B'_∞, the bounding streamline $A'B'_\infty$ becomes the real axis of the ζ'-plane. Along AB_∞, $\theta = 2\pi$ and the real part of ζ' varies from 0 to ∞. At I'_∞, $\zeta = -\infty$ and $\zeta' = \infty + i\pi$. II'_∞ plots as a horizontal line in the ζ'-plane extending from $(0,\pi)$ to (∞,π).

As the ζ'-plane shows the streamlines in the form of a semi-infinite strip, the Schwarz-Christoffel transformation of Sec. 86 can be employed. It has the form

$$\zeta' = C \cosh^{-1} t + D$$

where C and D are constants to be determined by the size and location of the figure in the ζ'-plane, where the two vertices A, A' have been selected at $t = -1$, $t = +1$, respectively, and where B_∞ has been selected at $t = -\infty$ in the t-plane.

Substituting two points into the equation to determine C and D: for A, $\zeta' = i2\pi$, $t = -1$, and for A', $\zeta' = 0$, $t = +1$. This results in

$$i2\pi = C \cosh^{-1}(-1) + D, \qquad 0 = C \cosh^{-1}(1) + D$$

From the second equation, since $\cosh^{-1}(1) = 0$, $D = 0$; hence, from the first equation

$$\cosh \frac{i2\pi}{C} = \cos \frac{2\pi}{C} = -1$$

and $C = 2$. The relation between ζ' and t is then

$$\zeta' = 2 \cosh^{-1} t$$

To locate the figure in the t-plane the coordinates of the end points in the ζ'-plane may be substituted into the equation. For B_∞,

$$\zeta' = \underset{x \to \infty}{L}\ x + i2\pi$$

hence,

$$t = \cosh \frac{\zeta'}{2} = \underset{x \to \infty}{L}\ \cosh\left(\frac{x}{2} + i\pi\right)$$

Expanding,

$$t = \underset{x \to \infty}{L}\ \left(\cosh \frac{x}{2} \cos \pi + i \sinh \frac{x}{2} \sin \pi\right) = -\infty$$

since the imaginary term is zero for all finite values of x and is assumed continuous in the limit as $x \to \infty$. For I'_∞

$$\zeta = \mathop{L}_{x \to \infty} x + i\pi$$

hence,

$$t = \mathop{L}_{x \to \infty} \cosh \left(\frac{x}{2} + i\frac{\pi}{2} \right) = i \infty$$

Since the streamline $I'_\infty I$ in the t-plane is not parallel to the real axis, this plane does not suffice as a w-plane and another transformation is required, this time from the upper half of the t-plane to an infinite strip in the w-plane. This last transformation, described in Sec. 87, is

$$w = C' \ln t + D' \tag{9}$$

where $w = \phi + i\psi$ and C', D' are to be determined from the location of axes in the w-plane and from the scale.

The streamline $I'_\infty I_\infty$ in the z-plane may be taken as $\psi = 0$. Letting the discharge per unit width of slot be $2b$, then along the streamline AI_∞, $\psi = b$; and along the streamline $A'I_\infty$, $\psi = -b$. Since the velocity at I_∞ is unity, the thickness of jet must be $2b$ there. Selecting $\phi = 0$ as the equipotential line through AA' in the z-plane, $\phi = \infty$ at I'_∞ and $\phi = -\infty$ at I_∞.

Letting $t = -1$, $w = ib$ for point A and $t = 1$, $w = -ib$ for point A', Eq. (9) yields

$$ib = C' \ln (-1) + D', \qquad -ib = C' \ln (1) + D'$$

From the second equation, since $\ln (1) = 0$, $D' = -ib$; and from the first equation,

$$\frac{2ib}{C'} = \ln (-1) = \ln e^{i\pi} = i\pi$$

and $C' = 2b/\pi$. The transformation becomes

$$w = \frac{2b}{\pi} \ln t - ib$$

To determine the location of the various points in the w-plane, substitutions can be made into this formula. For example, the line II'_∞ in the t-plane may be written $t = iy$, $0 \leqslant y < \infty$. Then

$$w = \frac{2b}{\pi} \ln iy - ib = \frac{2b}{\pi} \ln y$$

since $iy = ye^{i\pi/2}$. w is real and varies from $-\infty$ to $+\infty$ as y varies from 0 to ∞.

Collecting the transformations,

$$\zeta = -\frac{dz}{dw} = \left|\frac{1}{q}\right| e^{i\theta} = \frac{1}{q}\left(\frac{u}{q} + i\frac{v}{q}\right) \tag{10}$$

$$\zeta' = \ln \zeta = \ln \left|\frac{1}{q}\right| + i\theta \tag{11}$$

$$\zeta' = 2 \cosh^{-1} t \tag{12}$$

$$w = \frac{2b}{\pi} \ln t - ib \tag{13}$$

The variables ζ, ζ', t can be eliminated from these equations and w expressed as a function of z by integration of $\dfrac{dz}{dw}$. It is more convenient, however, to use the formulas in their present form.

The free streamlines are easily plotted by expressing x and y as parameters of θ. In the t-plane the free streamline $A'I$ is real and extends from $t = 1$ to $t = 0$. From Eqs. (11) and (12), since $q = 1$,

$$\zeta' = \ln \zeta = i\theta = 2 \cosh^{-1} t$$

or

$$t = \cosh \frac{i\theta}{2} = \cos \frac{\theta}{2}$$

From Eq. (13)

$$w = \phi + i\psi = \frac{2b}{\pi} \ln \cos \frac{\theta}{2} - ib$$

and

$$\phi = \frac{2b}{\pi} \ln \cos \frac{\theta}{2}$$

Along the free streamlines $-\dfrac{\partial \phi}{\partial s} = q = 1$; hence, ϕ may be replaced by $-s$, giving

$$s = \frac{2b}{\pi} \ln \sec \frac{\theta}{2}$$

where s is measured along $A'I$ from A'. This is the intrinsic equation of the curve; and by differentiation,

$$ds = \frac{2b}{\pi} \tan \frac{\theta}{2} d\left(\frac{\theta}{2}\right)$$

writing

$$ds = dx \sec \theta = dy \csc \theta$$

the parametric equations are obtained by integration. Thus,

$$x = \frac{2b}{\pi}\left(\sin^2\frac{\theta}{2} - \ln\sec\frac{\theta}{2}\right)$$

$$y = \frac{b}{\pi}(\theta - \sin\theta)$$

where the origin of the xy-system is at A'. The free streamline is described as θ varies from 0 to π. Since the asymptotic value of y is b, the distance between walls AA' is $4b$, and the coefficient of contraction is $\frac{1}{2}$, in agreement with Borda's theory. Figure 96 shows the free streamlines drawn to scale.

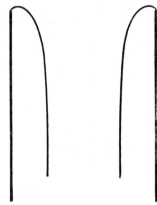

91. Flow out of a Two-dimensional Orifice. The solution for flow out of a slot in a flat plate is analytically similar to the preceding example, flow through Borda's mouthpiece. A large tank is assumed to have a rectangular slot of great length in the bottom, out of which fluid flows. To avoid infinite velocity at A and A', the flow separates from the boundary and leaves in a tangential direction. This example is not treated in as great detail as the preceding one; hence, reference to Sec. 90 should be

FIG. 96.—Free streamlines in Borda's mouthpiece in two dimensions.

made as needed. The same complex planes are used: z, ζ, ζ', t, and w, as shown in Fig. 97.

The problem is to find the shape of the free streamlines in the z-plane. In transforming the ζ-plane into the ζ'-plane the configuration is the same as in the previous example, except that the coordinates of the various significant points are changed. This requires a change in the constants in the transformation from ζ' to t. The general form is

$$\zeta' = C\cosh^{-1}t + D$$

The conditions to be substituted in are as follows: for A, $\zeta' = 0$, $t = -1$; for A', $\zeta' = -i\pi$, $t = 1$; thus,

$$0 = C\cosh^{-1}(-1) + D$$
$$-i\pi = C\cosh^{-1}(1) + D$$

from which

$$0 = i\pi C + D, \qquad -i\pi = D$$

therefore,

$$\zeta' = \cosh^{-1}t - i\pi \tag{14}$$

The transformation from the t-plane to the w-plane is exactly as before:

$$w = \frac{2b}{\pi} \ln t - ib \qquad (15)$$

where $2b$ is again the final thickness of the stream.   Along the free stream-

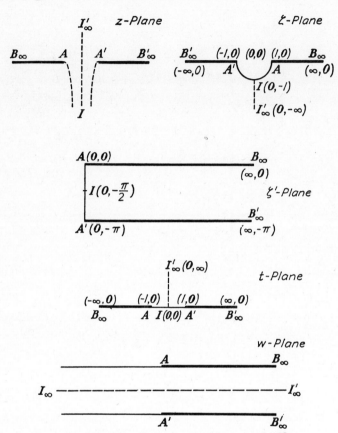

FIG. 97.—Mapping planes for a two-dimensional orifice.

line AI, t is real and varies from -1 to 0.   Since

$$\zeta = e^{i\theta}$$

along the free streamline

$$\zeta' = \ln e^{i\theta} = i\theta = \cosh^{-1} t - i\pi$$

and

$$t = \cosh i(\theta + \pi) = \cos (\theta + \pi), \qquad -\frac{\pi}{2} \leqslant \theta \leqslant 0$$

Substituting the value of t into Eq. (15),

$$w = \phi + i\psi = \frac{2b}{\pi} \ln \cos (\theta + \pi) - ib$$

from which, as before,

$$-\phi = s = \frac{2b}{\pi} \ln (\sec \theta)$$

and

$$ds = dx \sec \theta = dy \csc \theta = \frac{2b}{\pi} \tan \theta \, d\theta$$

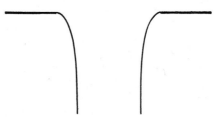

Fig. 98.—Free streamlines for flow out of a two-dimensional orifice.

Integrating,

$$x = \frac{2b}{\pi} \left(2 \sin^2 \frac{\theta}{2} - 1 \right) + C_1$$

$$y = \frac{2b}{\pi} \left[\ln \tan \left(\frac{\pi}{4} + \frac{\theta}{2} \right) - \sin \theta \right] + C_2$$

Taking the origin at A in the z-plane,

$$x = 0, \qquad y = 0, \qquad \theta = 0, \qquad C_1 = \frac{2b}{\pi}, \qquad C_2 = 0$$

The parametric equations become

$$\left. \begin{array}{l} x = \dfrac{4b}{\pi} \sin^2 \dfrac{\theta}{2}, \\[2mm] y = \dfrac{2b}{\pi} \left[\ln \tan \left(\dfrac{\pi}{4} + \dfrac{\theta}{2} \right) - \sin \theta \right] \end{array} \right\} -\frac{\pi}{2} \leqslant \theta \leqslant 0$$

The asymptotic value of x is $2b/\pi$ (for $\theta = -\pi/2$); hence, the total width of slot is $(4b/\pi) + 2b$. The coefficient of contraction is

$$\frac{2b}{(4b/\pi) + 2b} = \frac{\pi}{\pi + 2} = 0.611$$

The free streamlines are drawn to scale in Fig. 98. One of the curves may also represent the case of high velocity flow under a sluice gate when the line of symmetry is replaced by a solid boundary.

92. Infinite Stream Impinging upon a Fixed Plane Lamina. An infinite stream is divided by a rectangular fixed plane lamina of breadth l

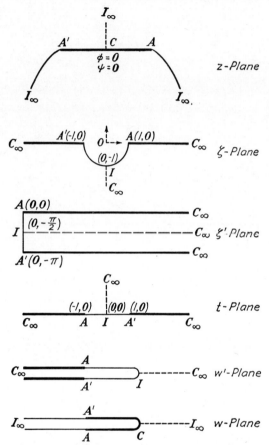

Fig. 99.—Mapping plane for an infinite stream impinging on a lamina.

at right angles to the stream. The flow separates into two portions divided internally by two free streamlines, as shown in Fig. 99. Unit velocity of the free streamlines is again assumed. The dividing streamline approaches the lamina along its bisector and has a stagnation point at the upstream point of contact. This streamline follows the lamina to its edges and then forms the free boundaries. The mapping planes for this transformation are shown in Fig. 99. The dividing streamline is taken as $\psi = 0$, and the equipotential line through the stagnation point as $\phi = 0$.

The transformations from z to ζ to ζ' to t are similar to those in the preceding examples. Equations (10) and (11) give the first two relationships, while the third one, from ζ' to t, is given by

$$\zeta' = \cosh^{-1} t - i\pi \tag{16}$$

Since both branches of the free streamline are $\psi = 0$, the region occupied by the fluid corresponds to the whole of the w-plane, which may be considered as bounded internally by a semi-infinite rectangle of zero thickness extending from $\phi = 0$ to $\phi = -\infty$.

In transforming from the t-plane to the w'-plane, the upper half of the t-plane is transformed into the *exterior* of the semi-infinite strip of zero thickness in the w'-plane. Taking a at $(t = -\infty)$, b at $I(t = 0)$, and d at $(t = +\infty)$, with $\beta = -\pi$, $\gamma = 0$, the Schwarz-Christoffel transformation gives

$$w' = -A \int \frac{dt}{t^{-1}} + B = -\frac{At^2}{2} + B$$

taking I at $w' = 0$, $B = 0$. Rewriting the constant A, the fourth transformation becomes

$$w' = -\frac{t^2}{C'}$$

Consideration of the transformation from the t- to the w'-plane shows that C' is a real positive constant. It is determined by the breadth of lamina l.

The final transformation to the w-plane is simply

$$w' = \frac{1}{w}$$

which moves I to the infinite regions and places C at the origin.

To determine the value of C', Eq. (16) is written

$$t = \cosh(\zeta' + i\pi) = -\frac{1}{2}\left(\zeta + \frac{1}{\zeta}\right)$$

since $\zeta' = \ln \zeta$. Along CA (z-plane), $\theta = 0$; therefore,

$$\zeta = \frac{1}{q}$$

and

$$t = -\frac{1}{2}\left(\frac{1}{q} + q\right), \qquad q = -t - \sqrt{t^2 - 1} \tag{17}$$

where the minus sign before the radical is selected so that $t = -\infty$ as $q \to 0$. Also along CA

$$\frac{\partial \phi}{\partial x} = -u = -q, \qquad \frac{dx}{d\phi} = -\frac{1}{q} \tag{18}$$

Integrating dx in the z-plane from C to A

$$\frac{l}{2} = \int_0^{l/2} dx = \int_{-\infty}^{-1} \frac{dx}{d\phi} \frac{d\phi}{dt} dt$$

taking the limits on the second integral by inspection of the t-plane (Fig. 99). Since

$$w = -\frac{C'}{t^2} = \phi + i\psi$$

for real values of t, $\psi = 0$ and

$$\frac{d\phi}{dt} = \frac{2C'}{t^3}$$

Substituting into the integral, using Eqs. (17) and (18),

$$\frac{l}{2} = -2C' \int_{-\infty}^{-1} \frac{dt}{qt^3} = -2C' \int_{-\infty}^{-1} (-t + \sqrt{t^2 - 1}) \frac{dt}{t^3}$$

This is easily integrated by making the change of variable $t = -1/t'$, thus,

$$\frac{l}{2} = 2C' \int_0^1 (1 + \sqrt{1 - t'^2}) \, dt' = 2C' + \frac{\pi C'}{2}$$

Hence, the relation between l and C' is

$$C' = \frac{l}{\pi + 4}$$

Along the free streamline $AI(q = 1)$

$$\zeta' = \ln \zeta = \ln e^{i\theta} = i\theta$$

and from Eq. (16)

$$t = \cosh(\ln \zeta + i\pi) = -\cos \theta$$

Since

$$w = \phi + i\psi = -\frac{C'}{t^2}$$

and t is real along AI,

$$\phi = -s = -\frac{C'}{t^2} = -C' \sec^2 \theta$$

The intrinsic equation of the free streamline is

$$s = \frac{l}{\pi + 4} \sec^2 \theta$$

where θ varies from 0 to $-\pi/2$ along AI. In parametric form this becomes

$$\left.\begin{aligned} x &= \frac{2l}{\pi + 4}\left(\sec \theta + \frac{\pi}{4}\right), \\ y &= \frac{l}{4\pi}\left[\sec \theta \tan \theta - \ln \tan \left(\frac{\pi}{4} + \frac{\theta}{2}\right)\right] \end{aligned}\right\} -\frac{\pi}{2} \leqslant \theta \leqslant 0$$

with the origin taken as the center of the lamina. The free streamlines are plotted in Fig. 100.

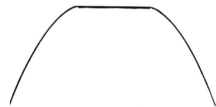

Fig. 100.—Free streamlines for a fluid impinging at right angles against a rectangular plane lamina.

The resultant force on the lamina can be found by applying the Bernoulli equation:

$$p + \tfrac{1}{2}\rho q^2 = p_0 + \tfrac{1}{2}\rho q_0^2$$

where p_0, q_0 refer to the pressure intensity and velocity of the undisturbed fluid. q_0 has been taken as unity, and the pressure intensity on the downstream side of the lamina is the same as that of the undisturbed fluid because the wake is assumed to be at rest. Then

$$p - p_0 = \frac{\rho}{2}(1 - q^2)$$

is the excess pressure intensity on the upstream face. The resultant force per unit length of lamina is

$$\begin{aligned} 2\int_0^{l/2} \frac{\rho}{2}(1 - q^2)\, dx &= \rho \int_{-\infty}^{-1}(1 - q^2)\frac{dx}{dt}\, dt \\ &= -2\rho C' \int_{-\infty}^{-1}\left(\frac{1}{q} - q\right)\frac{dt}{t^3} \\ &= -4\rho C' \int_{-\infty}^{-1} \sqrt{t^2 - 1}\, \frac{dt}{t^3} \\ &= \pi \rho C' \end{aligned}$$

This resultant force is for unit velocity. The Bernoulli equation shows that the resultant pressure varies as the square of the velocity of the stream. Therefore, the resultant force per unit length of lamina for any q_0 is

$$\pi \rho C' q_0{}^2 = \frac{\pi}{\pi + 4} \rho q_0{}^2 l = 0.440 \rho q_0{}^2 l$$

With actual fluids in which the wake has the same fluid as the stream, the assumption that the fluid in the wake is stationary is considerably in error. The pressure on the downstream side is actually less than that of the undisturbed stream.

Exercises

1. Map the line $y = l/2$ onto the t-plane of Fig. 92 using Eq. (4).

2. Find the mapping function for the semi-infinite rectangle of Fig. 92 for c at the origin in the t-plane, a, b, d remaining as in the figure.

$$Ans. \quad z = \frac{l}{\pi} \ln (2 \sqrt{t + t^2} + 2t + 1).$$

3. Locate the following points on the t-plane of Exercise 2:

(a) $z = il$　　　　　　　　　　(b) $z = l + i\dfrac{l}{2}$

(c) $z = 0$　　　　　　　　　　(d) $z = i\dfrac{l}{2}$

4. Plot the y-axis between $y = 0$ and $y = l$ on the t-plane for Fig. 93.

5. Complete the problem in Sec. 89, and plot the free streamlines. Find the coefficient of contraction for this opening.　　　　*Ans.* $C = 0.747$.

CHAPTER IX

VORTEX MOTION

An example of a two-dimensional vortex is given in Sec. 26 in which the vorticity, or region of rotational flow, is confined to a single line. In this chapter the theory is developed for vortex motion where the vorticity is confined to lines, tubes, and sheets or is unconfined, and examples are considered for both two-dimensional and three-dimensional flow.

THREE-DIMENSIONAL RELATIONSHIPS

93. Vorticity, Vortex Lines, and Vortex Tubes. The quantities ξ, η, ζ, defined as

$$
\left.
\begin{aligned}
\xi &= \frac{\partial w}{\partial y} - \frac{\partial v}{\partial z} \\[4pt]
\eta &= \frac{\partial u}{\partial z} - \frac{\partial w}{\partial x} \\[4pt]
\zeta &= \frac{\partial v}{\partial x} - \frac{\partial u}{\partial y}
\end{aligned}
\right\}
\tag{1}
$$

are known as the components of vorticity in the xyz-directions, respectively. Referring to Sec. 9, the vorticity components are seen to be twice the components of rotation about axes parallel to the xyz-axes, respectively. In the preceding chapters the flow has been assumed irrotational, except for isolated points and lines. Where ξ, η, and ζ have nonzero values throughout regions of the fluid, the flow is rotational. The theory of vortex motion was developed by Helmholz and was published in 1858.[1]

Continuous lines drawn in the fluid such that the vorticity vector (whose scalar components are ξ, η, ζ) is everywhere tangent to the lines are called *vortex lines*. The definition is analogous to that of streamlines (Sec. 16). The differential equations for vortex lines are

$$
\frac{dx}{\xi} = \frac{dy}{\eta} = \frac{dz}{\zeta}
\tag{2}
$$

The vortex lines passing through a small closed curve form the surface

---

[1] Helmholz, Ueber Integrale der hydrodynamischen Gleichungen welche den Wirbelbewegungen entsprechen, *Crelle's Journal*, Vol. 55, 1858; translated by Tait, *Phil. Mag.*, Vol. 33 (4th ser.), 485.

of a *vortex tube*. The fluid contained within this tube is referred to as a *vortex filament*, or merely as a *vortex*. In the following sections some of the properties of vortices are developed.

94. Circulation about a Vortex Tube. Letting l, m, n be the direction cosines of the normal to the surface of a vortex tube, the component of vorticity normal to it must be zero, or

$$l\xi + m\eta + n\zeta = 0 \tag{3}$$

Referring to Fig. 101, the line integral of the velocity around the path $ABCDD'C'B'A'A$ must be zero from Stokes' theorem (Sec. 24) as follows:

$$\int (u\,dx + v\,dy + w\,dz) = \int \int \left(\frac{\partial v}{\partial x} - \frac{\partial u}{\partial y}\right) dx\,dy$$
$$+ \int \int \left(\frac{\partial w}{\partial y} - \frac{\partial v}{\partial z}\right) dy\,dz$$
$$+ \int \int \left(\frac{\partial u}{\partial z} - \frac{\partial w}{\partial x}\right) dz\,dx$$
$$= \int (l\xi + m\eta + n\zeta)\,dA$$
$$= 0$$

since the integrand of the surface integral [Eq. (3)] is zero over the surface of the vortex tube. Rewriting the line integral,

$$\int_{ABCD} + \int_{DD'} + \int_{D'C'B'A'} + \int_{A'A} = 0$$

Since the two lines DD' and AA' can be made to approach each other, the sum of the line integrals along them cancel, leaving

$$\int_{ABCD} = \int_{A'B'C'D'}$$

which proves that the circulation around the vortex tube is the same at both sections. As the two circuits $ABCD$, $A'B'C'D'$ are any two circuits about a

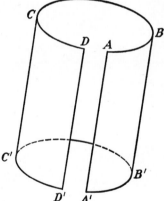

Fig. 101.—Circuit around a vortex tube.

vortex tube, the circulation about a vortex tube has thus been shown to be constant. Furthermore, from Sec. 25, the circulation about a small-area element is given by $2\omega_n\sigma$, where ω_n is the normal component of the rotation and σ is the elemental cross-sectional area of the vortex tube.

As the circulation is constant about a vortex tube, the vortex tube cannot end in the fluid but must either close upon itself or end at a fluid boundary. The "strength" of a vortex tube is its circulation, or the product of its vorticity and its cross-sectional area.

95. Vortex Lines Move with the Fluid. A theorem that the circulation of any circuit moving with the fluid is constant is first proved, and from this it can be shown that vortex lines move with the fluid.

Let A and B be the end points of a line drawn in the fluid such that each point on the line moves with the fluid particles. The rate at which flow along the line AB is increasing is to be calculated. Let δx, δy, δz be projections of an element of this line on the coordinate axes. Then

$$\frac{D(u\,\delta x)}{Dt} = \frac{Du}{Dt}\,\delta x + u\,\frac{D\,\delta x}{Dt} \tag{4}$$

where $\dfrac{Du}{Dt}$ is the acceleration in the x-direction, while $\dfrac{D\,\delta x}{Dt}$ is the rate that δx is increasing due to the fluid motion; *i.e.*, it is δu. Assuming that the extraneous forces have a single-valued potential and using Eqs. (3) and (16), Chap. II,

$$\frac{Du}{Dt} = -\frac{\partial\Omega}{\partial x} - \frac{1}{\rho}\frac{\partial p}{\partial x}$$

$$\frac{Dv}{Dt} = -\frac{\partial\Omega}{\partial y} - \frac{1}{\rho}\frac{\partial p}{\partial y}$$

$$\frac{Dw}{Dt} = -\frac{\partial\Omega}{\partial z} - \frac{1}{\rho}\frac{\partial p}{\partial z}$$

Then

$$\frac{D}{Dt}(u\,\delta x + v\,\delta y + w\,\delta z) = -\frac{\partial\Omega}{\partial x}\,\delta x - \frac{\partial\Omega}{\partial y}\,\delta y - \frac{\partial\Omega}{\partial z}\,\delta z$$

$$-\frac{1}{\rho}\left(\frac{\partial p}{\partial x}\,\delta x + \frac{\partial p}{\partial y}\,\delta y + \frac{\partial p}{\partial z}\,\delta z\right)$$

$$+ u\,\delta u + v\,\delta v + w\,\delta w$$

$$= -\delta\Omega - \frac{\delta p}{\rho} + \frac{\delta q^2}{2}$$

Integrating from A to B,

$$\frac{D}{Dt}\left[\int_A^B (u\,dx + v\,dy + w\,dz)\right] = \left[\frac{1}{2}q^2 - \Omega - \int\frac{dp}{\rho}\right]_A^B \tag{5}$$

if ρ is a function of p only.

Equation (5) gives the rate of increase of flow along the line AB.

If A and B coincide so that the line AB is a closed curve, Eq. (5) shows that the circulation around any closed curve moving with the fluid is constant for all time.   This proves that if irrotational flow exists within some portion of fluid, then the circulation about any closed curve is zero and remains zero; hence, the permanence of irrotational flow is established.

Any closed circuit drawn on the surface S of a vortex tube has circulation zero at time t.   Using the above theorem the circulation remains zero for the circuit now on the surface S_1, composed of the same particles at $t + \delta t$.   Therefore, the surface S_1 must also be a vortex tube.   This means that a vortex tube moves with the fluid.   The intersection of two vortex tubes must be along vortex lines; and hence, a vortex line must move with the fluid.

96. Determination of Velocity Components from Vorticity Components.   The velocity components at any point in space due to the presence of vortices in an incompressible fluid are obtained in the following manner:

The flow through all surfaces having a common boundary curve must be the same for an incompressible fluid.   The flow, therefore, depends on the boundary only, and it may be assumed that the flow is given by the line integral of some vector around the boundary curve.   In equation form

$$\int (lu + mv + nw)\, dS = \int (F\, dx + G\, dy + H\, dz) \tag{6}$$

where l, m, n are the direction cosines of the normal to the surface element dS; F, G, H are scalar components of some vector; and u, v, w are the scalar components of the velocity vector.   Using Stokes' theorem (Sec. 24),

$$\int (F\, dx + G\, dy + H\, dz) = \int \left[l\left(\frac{\partial H}{\partial y} - \frac{\partial G}{\partial z}\right) + m\left(\frac{\partial F}{\partial z} - \frac{\partial H}{\partial x}\right) \right.$$
$$\left. + n\left(\frac{\partial G}{\partial x} - \frac{\partial F}{\partial y}\right) \right] dS \tag{7}$$

Comparing Eqs. (6) and (7), it is obvious that

$$\left. \begin{aligned} u &= \frac{\partial H}{\partial y} - \frac{\partial G}{\partial z} \\ v &= \frac{\partial F}{\partial z} - \frac{\partial H}{\partial x} \\ w &= \frac{\partial G}{\partial x} - \frac{\partial F}{\partial y} \end{aligned} \right\} \tag{8}$$

Substituting the values of u, v, w from Eqs. (8) into Eqs. (1), the following equations are obtained:

$$\left.\begin{aligned}
\xi &= \frac{\partial}{\partial x}\left(\frac{\partial F}{\partial x} + \frac{\partial G}{\partial y} + \frac{\partial H}{\partial z}\right) - \nabla^2 F \\
\eta &= \frac{\partial}{\partial y}\left(\frac{\partial F}{\partial x} + \frac{\partial G}{\partial y} + \frac{\partial H}{\partial z}\right) - \nabla^2 G \\
\zeta &= \frac{\partial}{\partial z}\left(\frac{\partial F}{\partial x} + \frac{\partial G}{\partial y} + \frac{\partial H}{\partial z}\right) - \nabla^2 H
\end{aligned}\right\} \tag{9}$$

Values of F, G, H that satisfy Eqs. (9) may, by substitution into Eqs. (8), yield the desired velocity components. A particular solution is obtained by taking

$$\frac{\partial F}{\partial x} + \frac{\partial G}{\partial y} + \frac{\partial H}{\partial z} = 0 \tag{10}$$

and requiring the solution to satisfy Eq. (10) and

$$\nabla^2 F = -\xi, \qquad \nabla^2 G = -\eta, \qquad \nabla^2 H = -\zeta \tag{11}$$

These latter equations have the form of the Poisson equation.[1]

The solutions to Eqs. (10) and (11) are shown to be

$$\left.\begin{aligned}
F &= \frac{1}{4\pi}\iiint \frac{\xi'}{r}\, dx'\, dy'\, dz' \\
G &= \frac{1}{4\pi}\iiint \frac{\eta'}{r}\, dx'\, dy'\, dz' \\
H &= \frac{1}{4\pi}\iiint \frac{\zeta'}{r}\, dx'\, dy'\, dz'
\end{aligned}\right\} \tag{12}$$

The integrals are carried out over all space where the vorticity components have nonzero values; ξ', η', ζ' are the vorticity components of the volume element $dx'\, dy'\, dz'$ at (x',y',z'); F, G, H are evaluated for (x,y,z); and the distance from (x,y,z) to (x',y',z') is

$$r = \sqrt{(x - x')^2 + (y - y')^2 + (z - z')^2}$$

To prove that Eqs. (12) are solutions of Eqs. (11), the volume integrals are considered in two parts; thus

$$F = F_* + F'$$

where F_* represents all the space integral except a small sphere of radius r with center at (x,y,z), and F' represents the integral for the small sphere.

---

[1] E. B. Wilson, "Advanced Calculus," pp. 546–552, Ginn & Company, Boston, 1912.

Since r cannot go to zero in this region, F_* always has a finite integrand. Differentiation under the integral sign is, therefore, permissible. ξ' is a function of (x',y',z') and is therefore treated as a constant in the partial derivatives with respect to x, y, and z. Thus

$$\frac{\partial F_*}{\partial x} = \frac{1}{4\pi} \int \int \int \xi' \frac{\partial}{\partial x} \left(\frac{1}{r}\right) dx' \, dy' \, dz'$$

and

$$\frac{\partial^2 F_*}{\partial x^2} = \frac{1}{4\pi} \int \int \int \xi' \frac{\partial^2}{\partial x^2} \left(\frac{1}{r}\right) dx' \, dy' \, dz'$$

Hence,

$$\nabla^2 F_* = \frac{1}{4\pi} \int \int \int \xi' \, \nabla^2 \left(\frac{1}{r}\right) dx' \, dy' \, dz' = 0$$

since $\nabla^2(1/r)$ is identically zero.

The second space integral may be written

$$F' = \frac{1}{4\pi} \frac{\xi}{r} \delta V$$

where δV is the volume of the small sphere. As ξ' is assumed continuous, it takes on the value ξ at the center of the sphere and may be considered constant for the small sphere. Writing $\mu = \xi \, dV$,

$$F' = \frac{\mu}{4\pi r}$$

has the form of a velocity potential for source of strength μ at (x,y,z). Then,[1] considering the sphere fixed in space and size,

$$\frac{\partial F'}{\partial x} = \frac{dF'}{dr} \frac{\partial r}{\partial x} = -\frac{\mu}{4\pi r^2} \frac{x - x'}{r}$$

Substituting for μ and for the volume of the sphere,

$$\frac{\partial F'}{\partial x} = -\frac{\xi \frac{4}{3}\pi r^3}{4\pi r^2} \frac{x - x'}{r} = -\xi \frac{x - x'}{3}$$

Differentiating again,

$$\frac{\partial^2 F'}{\partial x^2} = -\frac{\xi}{3}$$

[1]
$$r^2 = (x - x')^2 + (y - y')^2 + (z - z')^2$$

$$2r \frac{\partial r}{\partial x} = 2(x - x')$$

and

$$\frac{\partial r}{\partial x} = \frac{x - x'}{r}$$

Similarly,

$$\frac{\partial^2 F'}{\partial y^2} = -\frac{\xi}{3}, \qquad \frac{\partial^2 F'}{\partial z^2} = -\frac{\xi}{3}$$

Adding,

$$\nabla^2 F' = -\xi$$

Hence,

$$\nabla^2 F = \nabla^2 F_* + \nabla^2 F' = -\xi$$

and the first of Eqs. (12) has been proved a solution of $\nabla^2 F = -\xi$. The other two equations are proved to be solutions in an analogous manner.

It must now be shown that Eq. (10) is satisfied by the solutions given in Eqs. (12). First

$$\frac{\partial F_*}{\partial x} = \frac{1}{4\pi} \int\int\int \zeta' \frac{\partial}{\partial x}\left(\frac{1}{r}\right) dx'\, dy'\, dz'$$

Since[1]

$$\frac{\partial}{\partial x} = -\frac{\partial}{\partial x'}$$

$$\frac{\partial F_*}{\partial x} = -\frac{1}{4\pi} \int\int\int \zeta' \frac{\partial}{\partial x'}\left(\frac{1}{r}\right) dx'\, dy'\, dz'$$

Integrating by parts

$$\frac{\partial F_*}{\partial x} = -\frac{1}{4\pi} \int\int \frac{\zeta'}{r}\, dy'\, dz' + \frac{1}{4\pi} \int\int\int \frac{1}{r}\frac{\partial \zeta'}{\partial x'}\, dx'\, dy'\, dz'$$

Hence,

$$\frac{\partial F_*}{\partial x} + \frac{\partial G_*}{\partial y} + \frac{\partial H_*}{\partial z} = -\frac{1}{4\pi} \int \frac{1}{r}(l\xi' + m\eta' + n\zeta')\, dS$$

$$+ \frac{1}{4\pi} \int\int\int \frac{1}{r}\left(\frac{\partial \xi'}{\partial x'} + \frac{\partial \eta'}{\partial y'} + \frac{\partial \zeta'}{\partial z'}\right) dx'\, dy'\, dz'$$

where l, m, n are the direction cosines of the normal to the boundary surface dS. Vortex filaments must either close or end at a boundary. If they end at a boundary, then both ends of the filament may be con-

[1]

$$\frac{\partial f(r)}{\partial x} = \frac{\partial f(r)}{\partial r}\frac{\partial r}{\partial x}, \qquad \frac{\partial f(r)}{\partial x'} = \frac{\partial f(r)}{\partial r}\frac{\partial r}{\partial x'},$$

but

$$\frac{\partial r}{\partial x} = \frac{x - x'}{r}, \qquad \frac{\partial r}{\partial x'} = -\frac{x - x'}{r}$$

hence

$$\frac{\partial f(r)}{\partial x} = -\frac{\partial f(r)}{\partial x'}$$

sidered to close in a greater space S' in which only re-entrant vortex filaments occur. Then over the boundary either $\xi' = \eta' = \zeta' = 0$ or

$$l\xi' + m\eta' + n\zeta' = 0$$

hence, the surface integral vanishes. Since

$$\frac{\partial \xi}{\partial x} + \frac{\partial \eta}{\partial y} + \frac{\partial \zeta}{\partial z} = 0$$

everywhere, by substitution of Eqs. (1), it has been proved that

$$\frac{\partial F_*}{\partial x} + \frac{\partial G_*}{\partial y} + \frac{\partial H_*}{\partial z} = 0$$

Since

$$\frac{\partial F'}{\partial x} + \frac{\partial G'}{\partial y} + \frac{\partial H'}{\partial z} = -\frac{1}{3}\left[\xi(x - x') + \eta(y - y') + \zeta(z - z')\right]$$

in the limit as the small radius approaches zero $x \to x'$, $y \to y'$, and $z \to z'$, or

$$\frac{\partial F'}{\partial x} + \frac{\partial G'}{\partial y} + \frac{\partial H'}{\partial z} = 0$$

proving that

$$\frac{\partial F}{\partial x} + \frac{\partial G}{\partial y} + \frac{\partial H}{\partial z} = 0$$

for F, G, and H given by Eqs. (12).

In order to gain an insight into Eqs. (8), the velocity components at (x,y,z) due to the volume element δV at (x',y',z') are obtained from Eqs. (8) and (12):

$$\left.\begin{aligned}
\delta u &= \frac{1}{4\pi}\left[\zeta'\frac{\partial}{\partial y}\left(\frac{1}{r}\right) - \eta'\frac{\partial}{\partial z}\left(\frac{1}{r}\right)\right]\partial V \\
\delta v &= \frac{1}{4\pi}\left[\xi'\frac{\partial}{\partial z}\left(\frac{1}{r}\right) - \zeta'\frac{\partial}{\partial x}\left(\frac{1}{r}\right)\right]\partial V \\
\delta w &= \frac{1}{4\pi}\left[\eta'\frac{\partial}{\partial x}\left(\frac{1}{r}\right) - \xi'\frac{\partial}{\partial y}\left(\frac{1}{r}\right)\right]\partial V
\end{aligned}\right\} \quad (13)$$

Taking the partial derivatives of $1/r$, these equations become

$$\left.\begin{aligned}
\delta u &= -\frac{1}{4\pi}\left[(y - y')\zeta' - (z - z')\eta'\right]\frac{\delta V}{r^3} \\
\delta v &= -\frac{1}{4\pi}\left[(z - z')\xi' - (x - x')\zeta'\right]\frac{\delta V}{r^3} \\
\delta w &= -\frac{1}{4\pi}\left[(x - x')\eta' - (y - y')\xi'\right]\frac{\delta V}{r^3}
\end{aligned}\right\} \quad (14)$$

Multiplying the first of these equations by $(x - x')$, the second by $(y - y')$, and the third by $(z - z')$ and adding,

$$(x - x')\, \delta u + (y - y')\, \delta v + (z - z')\, \delta w = 0 \qquad (15)$$

Similarly, multiplying the first of these equations by ξ', the second by η', and the third by ζ' and adding,

$$\xi'\, \delta u + \eta'\, \delta v + \zeta'\, \delta w = 0 \qquad (16)$$

Equation (15) shows that the velocity at (x,y,z) due to dV is at right angles to r, since $x - x'$, $y - y'$, $z - z'$ are proportional to the direction

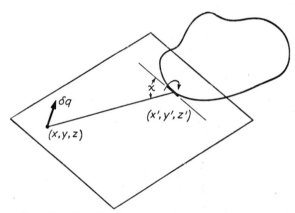

Fig. 102.—Velocity element due to segment of a vortex.

cosines of r and δu, δv, δw are proportional to the direction cosines of the velocity element. Likewise, Eq. (16) shows that the velocity element is at right angles to the axis of rotation of the element dV.

The magnitude of the velocity element may be written

$$\delta q = \sqrt{\delta u^2 + \delta v^2 + \delta w^2} = \frac{\delta V}{4\pi r^2}\, \omega'\, \sin \chi \qquad (17)$$

where ω' is the magnitude of the vorticity at (x',y',z'); *i.e.*,

$$\omega' = \sqrt{\xi'^2 + \eta'^2 + \zeta'^2}$$

and χ is the angle between r and the axis of rotation at (x',y',z'), as shown in Fig. 102. Equation (17) is easily proved to be correct by substituting for ω' and $\sin \chi$ where

$$\sin \chi = \sqrt{1 - \cos^2 \chi}$$

is used and $\cos \chi$ is expressed in terms of direction cosines.

Considering a segment of a vortex filament of length $\delta s'$ at (x',y',z') and having a strength κ,

$$\kappa = \omega' \sigma$$

where σ is the cross-sectional area of the filament. Then

$$\kappa \delta s' = \omega' \sigma \, \delta s' = \omega' \, \delta V$$

Equation (17) may now be written

$$\delta q = \frac{\kappa \delta s'}{4\pi} \frac{\sin \chi}{r^2} \tag{18}$$

The velocity element at (x,y,z) due to the segment of the vortex filament is perpendicular to the plane through (x,y,z) and the axis of rotation at (x',y',z').

97. Velocity Potential Due to a Vortex. In an incompressible fluid the velocity potential exists at points not on vortex filaments. The expression for velocity potential for a point external to a single reentrant vortex is obtained as follows:

From Eq. (13),

$$u = \frac{1}{4\pi} \int \left[\zeta' \frac{\partial}{\partial y} \left(\frac{1}{r}\right) - \eta' \frac{\partial}{\partial z} \left(\frac{1}{r}\right) \right] dV$$

writing $dV = \kappa ds'/\omega'$ and remembering that $\dfrac{\partial}{\partial y} = -\dfrac{\partial}{\partial y'}$ and that

$$\frac{\zeta'}{\omega'} = \frac{\delta z'}{\delta s'}, \; \cdots$$

$$u = \frac{\kappa}{4\pi} \int \left[\frac{\partial}{\partial z'} \left(\frac{1}{r}\right) dy' - \frac{\partial}{\partial y'} \left(\frac{1}{r}\right) dz' \right] \tag{19}$$

Using Stokes' theorem (Sec. 24) the line integral around the vortex filament can be replaced by a surface integral over any surface bounded by the vortex filament. Stokes' equation may be written

$$\int (P \, dx' + Q \, dy' + R \, dz') = \int \left[l \left(\frac{\partial R}{\partial y'} - \frac{\partial Q}{\partial z'} \right) \right.$$
$$\left. + m \left(\frac{\partial P}{\partial z'} - \frac{\partial R}{\partial x'} \right) + n \left(\frac{\partial Q}{\partial x'} - \frac{\partial P}{\partial y'} \right) \right] dS'$$

Letting

$$P = 0, \qquad Q = \frac{\partial}{\partial z'} \left(\frac{1}{r}\right), \qquad R = -\frac{\partial}{\partial y'} \left(\frac{1}{r}\right)$$

then

$$\frac{\partial R}{\partial y'} - \frac{\partial Q}{\partial z'} = -\frac{\partial^2}{\partial y'^2}\left(\frac{1}{r}\right) - \frac{\partial^2}{\partial z'^2}\left(\frac{1}{r}\right) = \frac{\partial^2}{\partial x'^2}\left(\frac{1}{r}\right)$$

from $\nabla^2(1/r) = 0$. Also

$$\frac{\partial P}{\partial z'} - \frac{\partial R}{\partial x'} = \frac{\partial^2}{\partial x' \, \partial y'}\left(\frac{1}{r}\right)$$

$$\frac{\partial Q}{\partial x'} - \frac{\partial R}{\partial y'} = \frac{\partial^2}{\partial x' \, \partial z'}\left(\frac{1}{r}\right)$$

Hence, Eq. (19) takes the form

$$u = \frac{\kappa}{4\pi}\int\left(l\frac{\partial}{\partial x'} + m\frac{\partial}{\partial y'} + n\frac{\partial}{\partial z'}\right)\frac{\partial}{\partial x'}\left(\frac{1}{r}\right)dS'$$

Since

$$\frac{\partial}{\partial x'} = -\frac{\partial}{\partial x}$$

and furthermore,

$$u = -\frac{\partial\phi}{\partial x}$$

then

$$\phi = \frac{\kappa}{4\pi}\int\left(l\frac{\partial}{\partial x'} + m\frac{\partial}{\partial y'} + n\frac{\partial}{\partial z'}\right)\left(\frac{1}{r}\right)dS'$$

which may be written[1]

$$\phi = \frac{\kappa}{4\pi}\int\frac{\cos\theta \, dS'}{r^2} \qquad (20)$$

where θ is the angle the normal to the surface element dS' makes with r, the line joining (x,y,z) to (x',y',z'). The integral in Eq. (20) may be interpreted as the solid angle subtended at (x,y,z) by the surface that is bounded by the vortex.

98. Vortex Sheets. A discontinuity in fluid velocity along a surface, such as the slippage of one layer of fluid over another, may be handled as a vortex sheet in an otherwise continuous flow. The normal components of velocity on the two sides of the discontinuity are assumed equal; *i.e.*,

$$lu + mv + nw = lu' + mv' + nw'$$

[1] $\quad l\dfrac{\partial}{\partial x'}\left(\dfrac{1}{r}\right) = l\dfrac{\partial}{\partial r}\left(\dfrac{1}{r}\right)\dfrac{\partial r}{\partial x'} = l\left(-\dfrac{1}{r^2}\right)\dfrac{(x'-x)}{r} = \dfrac{l}{r^2}\dfrac{(x-x')}{r}$

and

$$\cos\theta = l\frac{(x-x')}{r} + m\frac{(y-y')}{r} + n\frac{(z-z')}{r}$$

where l, m, n are the direction cosines of the normal to the surface and where u, v, w, and u', v', w' represent the velocity components on the two sides of the surface. Rewriting,

$$l(u - u') + m(v - v') + n(w - w') = 0 \qquad (21)$$

The components $u - u'$, $v - v'$, $w - w'$ represent the relative velocity, which is obviously tangent to the surface of discontinuity, since the components are proportional to the direction cosines of the vector.

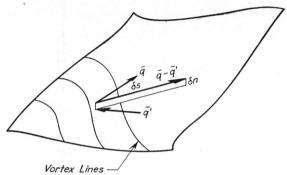

Vortex Lines

FIG. 103.—Circulation about element of vortex sheet.

Similarly, letting q, q' be the magnitudes of the velocity vector on either side of the discontinuity, the magnitude of the vector $\bar{q} - \bar{q}'$ is given by

$$|\bar{q} - \bar{q}'| = \sqrt{(u - u')^2 + (v - v')^2 + (w - w')^2}$$

Constructing a small rectangle with sides δs in the direction of the relative velocity and sides δn much smaller than δs, as shown in Fig. 103, the circulation is given by

$$\kappa = \omega'\sigma = |(\bar{q} - \bar{q}')| \, \delta s \qquad (22)$$

where ω' is the vorticity and σ the area of the rectangle $\delta n \, \delta s$. Rewriting,

$$\omega' \, \delta n = |\bar{q} - \bar{q}'|$$

showing that the vorticity varies inversely with the thickness δn for a given discontinuity. The layer, of thickness δn, is called a *vortex sheet*. It may be of infinitesimal thickness, in which case the vorticity is infinite, or it may have finite thickness, with finite vorticity; hence, the distribution of vorticity in a sheet in a fluid gives rise to a discontinuity in the velocity components tangent to the sheet.

A simple example is given by a uniform plane vortex sheet comprising the xy-plane with the vortex lines parallel to Ox, as in Fig. 104, and with a constant strength κ per unit length in direction Oy. As the strength is equal to the circulation, consider the element of length δy; let v be the velocity of the element on the side $z > 0$ and v' the velocity on the side $z < 0$. Then from Eq. (22) the circulation is

$$\kappa \, \delta y = (v' - v) \, \delta y$$

where the positive sense of rotation is from Oy to Oz.

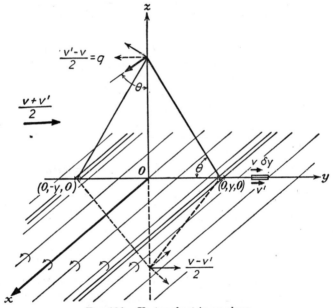

Fig. 104.—Vortex sheet in xy-plane.

To find the resultant velocity at any point, say $(0,0,z)$, due to the vortex sheet, the velocity due to an elemental strip of the vortex of breadth δy at y is $\delta q = \kappa \, \delta y / 2\pi r$ as in Sec. 26, where $r = \sqrt{y^2 + z^2}$. Taking another strip at $-y$, the resultant of the two strips is a velocity $2\delta q \sin \theta$ in the direction of the negative y-axis. Summing up the velocity contribution at $(0,0,z)$ due to all strips

$$q = \int 2 \sin \theta \, dq = \frac{\kappa}{2\pi} \int_{-\infty}^{\infty} \frac{z \, dy}{z^2 + y^2} = \frac{\kappa}{2}$$

since $\sin \theta = z/\sqrt{y^2 + z^2}$. When z is negative, the resultant velocity is reversed in direction. Superposing the uniform velocity $(v + v')/2$ parallel to Oy, the resulting flow is two uniform velocities v, v', respectively, as z is positive or negative.

99. Circular Vortex Rings.   Referring to Eq. (18)

$$\delta q = \frac{\kappa}{4\pi} \delta s' \frac{\sin \chi}{r^2} \qquad (18)$$

where δq is the speed at (x,y,z) due to the element of vortex of length $\delta s'$ at (x',y',z') of strength κ.   The distance between (x,y,z) and (x',y',z') is r, and χ is the angle between r and $\delta s'$ as shown in Fig. 105.   To find the

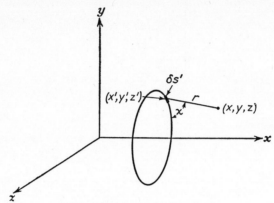

Fig. 105.—Circular vortex ring.

velocity at any point in space due to a circular vortex ring requires, in general, the use of elliptic integrals.   Several general conclusions regarding the motion of circular vortex rings having the same axis of symmetry, say the x-axis, may be drawn from Eq. (18).

Consider first a single vortex ring, lying in a plane parallel to the yz-plane, with the x-axis as its axis.   The velocity at any point on the ring due to any element δs of the ring is parallel to the x-axis, as it is normal to the plane through r and δs.   Hence, the velocity of the ring is parallel to the axis, and the radius does not change with time.   For a given strength κ the ring will translate more rapidly with a smaller radius; and as the radius becomes very large, the vortex ring tends to come to rest.

The velocity of fluid along the axis can be computed from Eq. (18). For convenience assume the circular vortex ring in the yz-plane and of radius r_0.   The speed at point $(x,0,0)$ is then given by

$$u = \frac{\kappa}{2} \frac{r_0^2}{(r_0^2 + x^2)^{3/2}}$$

since $\chi = \pi/2$, $r = \sqrt{r_0{}^2 + x^2}$. The direction is parallel to the x-axis. At the center of the ring the speed is $\kappa/2r_0$. Fluid near the ring is carried around the ring.

When two vortices have a distance between them, large compared with their radii, the effect of one on the other is negligible. When two vortex rings having the same axis are close to each other, each is affected both by its own field and by the field of the other. Due to the symmetry, both vortices remain circular. With rotation in the same direction, as

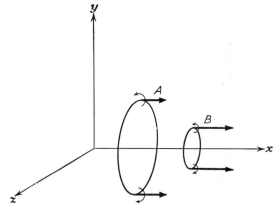

Fig. 106.—Two circular vortex rings with centers on x-axis.

in Fig. 106, the smaller ring will increase in size and the larger one will decrease in size. The velocity of vortex A will increase in the positive x-direction and vortex B will decrease in the same direction, so that vortex A will become smaller than B and pass through B. After ring A passes through B, it will start to grow and B will start to shrink, so that B will pass through A. This will continue indefinitely.

When two vortices have opposite rotations, equal strength, the same size, and rotation in the sense such that they tend to approach each other, they will grow larger and larger as they approach. Due to symmetry the plane midway between the two has no flow across it. It may then be taken as a solid wall, and the case of a single circular vortex approaching a wall is obtained.

TWO-DIMENSIONAL RELATIONSHIPS—RECTILINEAR VORTICES

In two-dimensional flow all vortex filaments must be straight and parallel. The computation of resulting fluid motions is therefore much simpler than for three-dimensional motion. Some general relationships valid for two-dimensional flow are derived, and then special cases are

considered.   In Secs. 100 to 103, inclusive, the area of the vortex filaments are taken as infinitesimal.

100. Velocity Due to Vortex System.   Referring to Fig. 107, the
velocity components at $P(x,y)$ due to the line vortex of strength κ_1 at
(x_1,y_1) are

$$u = -\frac{\kappa_1}{2\pi}\frac{y - y_1}{r_1{}^2}, \qquad v = \frac{\kappa_1}{2\pi}\frac{x - x_1}{r_1{}^2}$$

where the strength is defined as in Sec. 56 and is considered positive when
the sense of rotation is counterclockwise.   The velocity components at P
due to any number of vortices, n are

$$u = -\sum_{s=1}^{s=n}\frac{\kappa_s}{2\pi}\frac{y - y_s}{r_s{}^2}, \qquad v = \sum_{s=1}^{s=n}\frac{\kappa_s}{2\pi}\frac{x - x_s}{r_s{}^2} \tag{23}$$

Letting u_s, v_s denote the velocity components of a filament of strength
κ_s, the sum of the products $\kappa_s u_s$,
$\kappa_s v_s$ for any number of vortices is
easily shown to be zero, since pairs
of terms of opposite signs occur
such as

$$\kappa_1\frac{\kappa_3}{2\pi}\frac{y_1 - y_3}{r^2}, \qquad \kappa_3\frac{\kappa_1}{2\pi}\frac{y_3 - y_1}{r^2}$$

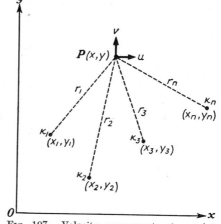

Considering κ_s as a mass, plus or
minus, the mass center of the
system is seen to be at rest.   The
mass center is given by

$$\bar{x} = \frac{\Sigma\kappa_s x_s}{\Sigma\kappa_s}, \qquad \bar{y} = \frac{\Sigma\kappa_s y_s}{\Sigma\kappa_s} \tag{24}$$

**101. Examples of Simple Vor-
tex Systems.**   Several s p e c i a l

Fig. 107.—Velocity components at a point
due to vortex system.

cases consisting of a finite number of vortices are considered in the
following examples:

a. Single Vortex.   From Eqs. (23) and (24) it is observed that the
vortex filament remains at rest.   The velocity of any point $P(x,y)$ is
given by

$$u = -\frac{\kappa}{2\pi}\frac{y - y_0}{r^2}, \qquad v = \frac{\kappa}{2\pi}\frac{x - x_0}{r^2}$$

where the vortex is at (x_0, y_0).   The speed is

$$q = \frac{\kappa}{2\pi r}$$

as in Sec. 56.

b. *Two Vortices of Equal Strength and the Same Sense.*   Let the strength of each vortex be κ and the positions be (x_1, y_1) and (x_2, y_2), as in

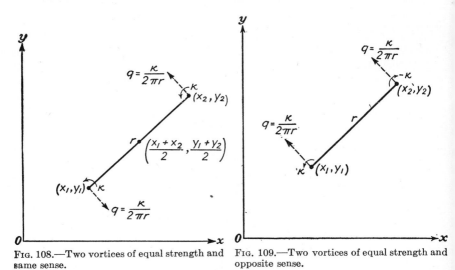

FIG. 108.—Two vortices of equal strength and same sense.

FIG. 109.—Two vortices of equal strength and opposite sense.

Fig. 108.   From Eqs. (24) the mass center at

$$\bar{x} = \frac{x_1 + x_2}{2}, \qquad \bar{y} = \frac{y_1 + y_2}{2}$$

remains fixed.   This is the mid-point between the two vortices.   Each vortex rotates around this mid-point with a speed $\kappa/2\pi r$.

c. *Two Vortices of Equal Strength and Opposite Sense.*   From Eqs. (24) the mass center of this system is shown to be at infinity.   From Fig. 109 the two filaments both have the same velocity; hence, the system translates with speed $\kappa/2\pi r$.   Since the plane bisecting the two points has no cross flow, it may be taken as a wall, and the case of a vortex filament moving parallel to a wall is obtained.   The speed of the vortex with respect to the wall is $\kappa/2\pi r$.

If the limiting case of two equal opposite vortices is taken as $r \to 0$, such that the product $\kappa r = \mu$ remains constant, a vortex doublet is obtained.   This limiting process is handled the same as the doublet of Sec. 61.

d. Two Vortices of Different Strengths. The center of mass of two unequal vortices at A and B (Fig. 110) lies along the line through AB. When κ_1, κ_2 have the same sign, the point of rotation is between A and B; but when their signs are different, it is on AB extended. Each vortex has a velocity due to the other; and since this is always at right angles to the line connecting them, they remain the fixed distance AB apart. Rotation centers are shown in Fig. 111 for two cases.

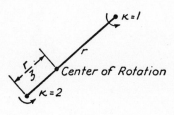

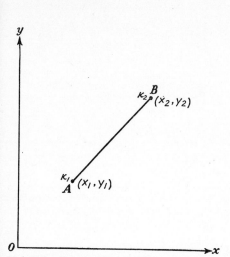

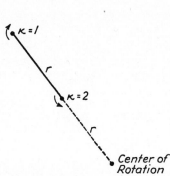

FIG. 110.—Two vortices of different strengths.   FIG. 111.—Centers of rotation for vortex pairs.

102. Infinite Row of Equal Vortices. The complex potential for a vortex of strength κ at $z = z_1$ is given by

$$w = \frac{\kappa}{2\pi} i \ln (z - z_1)$$

as in Sec. 56. For a series of $2n + 1$ vortices of the same strength κ spaced along the x-axis at a distance a apart with the center one at the origin,

$$w = \frac{i\kappa}{2\pi} \ln z(z - a)(z + a)(z - 2a)(z + 2a) \cdots (z - na)(z + na)$$

Rewriting,

$$w = \frac{i\kappa}{2\pi} \ln \frac{\pi z}{a} \left(1 - \frac{z^2}{a^2}\right)\left(1 - \frac{z^2}{4a^2}\right) \cdots \left(1 - \frac{z^2}{n^2 a^2}\right) + \text{constant}$$

When $n \to \infty$, this may be written

$$w = \frac{i\kappa}{2\pi} \ln \sin \frac{\pi z}{a} \tag{25}$$

neglecting the constant terms, which have no bearing on the velocity.

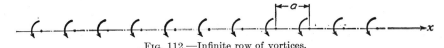

Fig. 112.—Infinite row of vortices.

The velocity can be obtained from the complex velocity

$$-\frac{dw}{dz} = u - iv = -\frac{i\kappa}{2a} \cot \frac{\pi z}{a}$$

Separating into real and pure imaginary parts

$$\left.\begin{array}{l} u = -\dfrac{\kappa}{2a} \dfrac{\sinh \dfrac{2\pi y}{a}}{\cosh \dfrac{2\pi y}{a} - \cos \dfrac{2\pi x}{a}} \\[4ex] v = \dfrac{\kappa}{2a} \dfrac{\sin \dfrac{2\pi x}{a}}{\cosh \dfrac{2\pi y}{a} - \cos \dfrac{2\pi x}{a}} \end{array}\right\} \tag{26}$$

Consideration of the effect of pairs of vortices (Fig. 112) at equal distances from a given vortex shows that the vortices remain fixed.

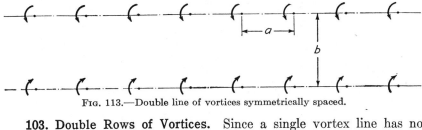

Fig. 113.—Double line of vortices symmetrically spaced.

103. Double Rows of Vortices. Since a single vortex line has no influence on the position of its own vortices, the motion of an individual vortex in a double row is found by determining the velocity caused by the second vortex line. With a symmetrical location, as in Fig. 113, with spacing a between vortices in a row and distance b between rows, the resultant velocity may be seen to be parallel to the rows. Referring to Eqs. (26) the translation velocity is given by letting $x = na$, $y = -b$:

$$U = \frac{\kappa}{2a} \coth \frac{\pi b}{a}, \qquad V = 0 \tag{27}$$

When one of the vortex rows is displaced parallel to itself $a/2$, such that the vortex of one row is opposite the mid-point between vortices

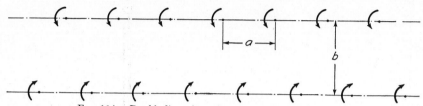

Fig. 114.—Double line of vortices unsymmetrically spaced.

of the other rows, as in Fig. 114, the motion of the vortices is given by substitution in Eqs. (26) for $x = (n + \frac{1}{2})a$, $y = -b$; thus

$$U = \frac{\kappa}{2a} \tanh \frac{\pi b}{a}, \qquad V = 0 \tag{28}$$

Consideration of pairs of vortices equally spaced from a vortex in the opposite row also shows that the motion is parallel to the rows.

The unsymmetrical case shown in Fig. 114 is called the Kármán vortex street after Th. von Kármán who pointed out that such a configuration arises when a body, such as a cylinder, moves through a fluid. He also showed that when $b/a = 0.281$, the motion is stable. The question of stability is fully discussed in Lamb's "Hydrodynamics," Art. 156, 1932.

104. Rectilinear Vortices with Finite Sections. For two-dimensional vortex motion with flow lines in planes parallel to the xy-plane

$$w = 0, \qquad \frac{\partial u}{\partial z} = 0, \qquad \frac{\partial v}{\partial z} = 0, \qquad \xi = 0, \qquad \eta = 0$$

and from Eqs. (1)

$$\zeta = \frac{\partial v}{\partial x} - \frac{\partial u}{\partial y} \tag{29}$$

From the equation for streamlines

$$v \, dx - u \, dy = 0$$

and from the equation of continuity

$$\frac{\partial u}{\partial x} + \frac{\partial v}{\partial y} = 0$$

the streamline equation is shown to be a perfect differential $d\psi$; therefore,

$$u = -\frac{\partial \psi}{\partial y}, \qquad v = \frac{\partial \psi}{\partial x}$$

and from Eq. (29)

$$\frac{\partial^2 \psi}{\partial x^2} + \frac{\partial^2 \psi}{\partial y^2} = \zeta \qquad (30)$$

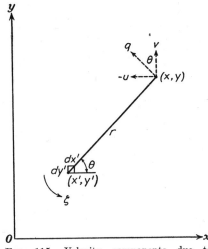

with the streamlines given by $\psi =$ constant. For those regions where $\zeta = 0$, the flow is irrotational and a velocity potential exists.

The velocity components at any point (x,y) due to a vortex filament at (x',y') can be obtained by reference to Fig. 115. The circulation around the v o r t e x f i l a m e n t of area $dx'\,dy'$ is $\kappa = \zeta'\,dx'\,dy'$, where ζ' is the vorticity at (x',y'). The speed q at (x,y) is then $\kappa/2\pi r$, and the velocity components

$$u = -q \sin \theta, \qquad v = q \cos \theta$$

Fig. 115.—Velocity components due to vorticity.

Expressing θ in terms of the coordinates,

$$\left.\begin{aligned} u &= -\frac{\partial \psi}{\partial y} = -\frac{\zeta'\,dx'\,dy'}{2\pi r}\frac{y - y'}{r} = -\frac{\zeta'\,dx'\,dy'}{2\pi}\frac{y - y'}{r^2} \\ v &= \frac{\partial \psi}{\partial x} = \frac{\zeta'\,dx'\,dy'}{2\pi r}\frac{x - x'}{r} = \frac{\zeta'\,dx'\,dy'}{2\pi}\frac{x - x'}{r^2} \end{aligned}\right\} \qquad (31)$$

The velocity components at (x,y) due to distributed vorticity are

$$\left.\begin{aligned} u &= -\frac{\partial \psi}{\partial y} = -\frac{1}{2\pi} \int\!\!\int \frac{\zeta'}{r^2}\,(y - y')\,dx'\,dy' \\ v &= \frac{\partial \psi}{\partial x} = \frac{1}{2\pi} \int\!\!\int \frac{\zeta'}{r^2}\,(x - x')\,dx'\,dy' \end{aligned}\right\} \qquad (32)$$

where the integration is carried out for the areas having nonzero values of ζ'. Integrating Eqs. (32),

$$\psi = \frac{1}{2\pi} \int\!\!\int \zeta' \ln r \, dx'\,dy' + \psi_0 \qquad (33)$$

which is easily verified by differentiation, remembering that

$$\frac{\partial \psi}{\partial x} = \frac{\partial \psi}{\partial r} \frac{\partial r}{\partial x}, \qquad r^2 = (x - x')^2 + (y - y')^2, \qquad \frac{\partial r}{\partial x} = \frac{x - x'}{r}, \text{ etc.}$$

ψ_0 can be selected to satisfy boundary equations.   For an unlimited fluid at rest at infinity it is a constant.

For a single vortex, where ψ is a function of r only and r is the distance from the vortex, from Eq. (30)

$$\frac{\partial^2 \psi}{\partial r^2} + \frac{1}{r} \frac{\partial \psi}{\partial r} = \zeta \tag{34}$$

This is obtained by change of independent variable, using $r^2 = x^2 + y^2$. The steps are as follows:

$$\frac{\partial \psi}{\partial x} = \frac{\partial \psi}{\partial r} \frac{\partial r}{\partial x}$$

$$\frac{\partial^2 \psi}{\partial x^2} = \frac{\partial^2 r}{\partial x^2} \frac{\partial \psi}{\partial r} + \frac{\partial^2 \psi}{\partial r^2} \left(\frac{\partial r}{\partial x}\right)^2$$

$$\frac{\partial r}{\partial x} = \frac{x}{r}$$

$$\frac{\partial^2 r}{\partial x^2} = \frac{r^2 - x^2}{r^3}, \text{ etc.}$$

105. Single Rectilinear Vortex with Finite Circular Section.   Let the vorticity be constant and equal to ζ throughout a circular area of radius a.   The stream functions are

$$\frac{\partial^2 \psi}{\partial x^2} + \frac{\partial^2 \psi}{\partial y^2} = \zeta \qquad \text{for } r < a$$

and

$$\frac{d^2 \psi}{dx^2} + \frac{d^2 \psi}{dy^2} = 0 \qquad \text{for } r > a$$

Expressed in terms of r,

$$\frac{d^2 \psi}{dr^2} + \frac{1}{r} \frac{d\psi}{dr} = \zeta \qquad \text{for } r < a \tag{35}$$

and

$$\frac{d^2 \psi}{dr^2} + \frac{1}{r} \frac{d\psi}{dr} = 0 \qquad \text{for } r > a \tag{36}$$

Equation (36) becomes, when integrated,

$$\psi = A \ln r + B$$

A particular integral of Eq. (35) is

$$\psi = \frac{\zeta r^2}{4}$$

The solutions for Eqs. (35) and (36) then become

$$\psi = A \ln r + B + \frac{\zeta r^2}{4} \qquad \text{for } r < a$$
$$\psi = C \ln r + D \qquad \text{for } r > a$$

The stream function should be finite for $r = 0$; hence, $A = 0$. For the motion to be continuous at the periphery of the vortex, ψ and $\partial\psi/\partial r$ must be continuous at $r = a$; thus

$$B + \frac{\zeta a^2}{4} = C \ln a + D$$

and

$$\frac{\zeta a}{2} = \frac{C}{a}$$

Eliminating the constants of integration and neglecting the additive constants,

$$\psi = -\frac{\zeta}{4}(a^2 - r^2) \qquad \text{for } r < a \tag{37}$$

and

$$\psi = \frac{\zeta a^2}{2} \ln \frac{r}{a} \qquad \text{for } r > a \tag{38}$$

It is evident from Eqs. (37) and (38) that the velocity is wholly tangential, having as its values $\zeta r/2$ inside the vortex and $\zeta a^2/2r$ outside the vortex. Outside the vortex a velocity potential exists. It may be found from Eq. (38) using the relations in Exercise 1, Chap. III; thus,

$$\phi = -\frac{\zeta a^2}{2}\theta \tag{39}$$

The complex potential, from Eqs. (38) and (39), is

$$w = i\frac{\zeta}{2}a^2 \ln \frac{z}{a} \tag{40}$$

Letting κ be the circulation, since $\kappa = \pi a^2 \zeta$,

$$\phi = -\frac{\kappa\theta}{2\pi}, \qquad \psi = \frac{\kappa}{2\pi}\ln\frac{r}{a} \tag{41}$$

106. Rankine's Combined Vortex. Considering the case of a rectilinear vortex of finite circular section as in Sec. 105, with axis vertical and subject to the action of gravity, a three-dimensional case is obtained, having a free surface. The expressions for velocity obtained in Sec. 105 are valid. Referring to Eq. (21), Sec. 10, for a given horizontal plane,

$$\frac{q^2}{2} - \frac{\partial \phi}{\partial t} + \frac{p}{\rho} = F(t)$$

is the Bernoulli equation for unsteady flow, where p is the absolute pressure for this application.

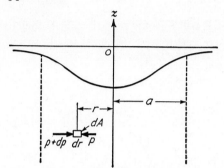

FIG. 116.—Rankine's combined vortex.

Applying the Bernoulli equation to the fluid outside the vortex, substituting $q = \kappa/2\pi r$, and simplifying for steady flow,

$$\frac{\kappa^2}{8\pi^2 r^2} + \frac{p}{\rho} = \frac{p'_\infty}{\rho} \qquad \text{for } r > a \tag{42}$$

where p'_∞ is the pressure at $r = \infty$ in the same horizontal plane for which p is determined. Referring to Fig. 116, with z measured vertically upward, the pressure at any point (r,z) is given by

$$\frac{p}{\rho} = \frac{p_\infty}{\rho} - \frac{\kappa^2}{8\pi^2 r^2} - gz \qquad \text{for } r > a \tag{43}$$

where p_∞ is the pressure at $z = 0$, $r = \infty$.

To find the pressure distribution inside the vortex, the equation of motion for radial direction is written from Fig. 116:

$$p \, dA - (p + dp) \, dA = \rho \, dA \, dr \left(-\frac{q^2}{r} \right) = -\rho \, dA \, dr \frac{\zeta^2}{4} r$$

or

$$dp = \rho \frac{\zeta^2}{4} r \, dr$$

Integrating,

$$p = \rho \frac{\zeta^2}{8} r^2 + P \qquad \text{for } r < a \tag{44}$$

where P is the pressure at $r = 0$ in the same horizontal plane as p. Since the pressure must be the same at $r = a$ in Eqs. (42) and (44), using $\kappa = \zeta \pi a^2$,

$$\frac{p'_\infty}{\rho} = \frac{P}{\rho} + \frac{\kappa^2}{4\pi^2 a^2}$$

Substituting in Eq. (44),

$$\frac{p}{\rho} = \frac{\kappa^2 r^2}{8\pi^2 a^4} + \frac{p'_\infty}{\rho} - \frac{\kappa^2}{4\pi^2 a^2}$$

If $p'_\infty < \kappa^2 \rho / 4\pi^2 a^2$, P will become negative for sufficiently small values of $r < a$. This means that a void will exist inside the vortex.

The pressure inside the vortex, as a function of z and r, is given by

$$\frac{p}{\rho} = C + \frac{\kappa^2 r^2}{8\pi^2 a^4} - gz, \qquad r < a \tag{45}$$

Letting p equal a constant to obtain the equations for free surface,

$$z = \frac{\kappa^2}{8\pi^2 a^4 g}\left(a^2 - \frac{a^4}{r^2}\right) + C, \qquad r > a \tag{46}$$

$$z = \frac{\kappa^2}{8\pi^2 a^4 g}(r^2 - a^2) + C, \qquad r < a \tag{47}$$

where the constants have been arranged so that $z = C$ when $r = a$. With the plane of the origin at the free surface at infinity, $z = 0$, $r = \infty$,

$$C = \frac{-\kappa^2}{8\pi^2 a^2 g}$$

The depth of depression at the axis is, from Eq. (47),

$$z = -\frac{\kappa^2}{4\pi^2 a^2 g}$$

A vortex of this type can be set up by allowing flow out of the bottom of a reservoir until the whirling motion is established; then by closing the opening, the center portion fills with liquid rotating as if it were a solid.

Exercises

1. Using Eq. (18), show that the velocity at (x,y,z) due to a rectilinear vortex of infinite length has the magnitude

$$q = \frac{\kappa}{2\pi r_0}$$

and that its direction is normal to the plane through the vortex and the point (x,y,z). The quantity r_0 is the perpendicular distance from (x,y,z) to the vortex.

2. A circular vortex ring of strength κ and radius a is located in the yz-plane with center of the ring at the origin. Find the velocity along the x-axis in terms of κ, a,

and x. $Ans.\quad q = \dfrac{\kappa a^2}{2(a^2 + x^2)^{\frac{3}{2}}}.$

3. Two parallel vortex rings normal to the x-axis and having centers on it are located in the planes $x = 0$ and $x = 3$ ft at a given instant. The ring in plane $x = 0$ has a radius 2 ft and circulation 6 ft^2 per sec in the sense that causes the other vortex ring to contract. The vortex ring at $x = 3$ ft, with radius of 1 ft, has a circulation 10 ft^2 per sec in the opposite sense. Find the velocity u along the x-axis.

$$Ans.\quad u = \frac{5}{(x^2 - 6x + 10)^{\frac{3}{2}}} - \frac{12}{(4 + x^2)^{\frac{3}{2}}}.$$

4. Find the rate at which the radius of the vortex at $x = 3$ ft (of Exercise 3) reduces.   *Ans.*   0.08 ft per sec.

5. Two two-dimensional vortices of strength 8 ft^2 per sec are located at $(0,0)$ and $(0,3)$ at a given instant.

(*a*) Find their location after 1.0 sec.

(*b*) If the sense of the vortex at the origin is reversed, find their positions after 1.0 sec.   *Ans.*   (*a*) $(0.42,0.06)$, $(-0.42,2.94)$; (*b*) $(0.425,0)$, $(0.425,3)$.

6. Find the centers of rotation of the following vortex pairs:

(*a*) $\kappa_1 = 3$ at $(1,1)$, $\kappa_2 = -4$ at $(2,3)$

(*b*) $\kappa_1 = 10$ at $(0,0)$, $\kappa_2 = 3$ at $(0,5)$

$Ans.$ (*a*) $(5,9)$; (*b*) $(0,\tfrac{15}{13})$.

7. An infinite row of equal vortices of strength 2 ft^2 per sec are uniformly spaced unit distance apart along the x-axis with one at the origin. If the one at the origin is displaced a very small distance to $(\Delta x, \Delta y)$, find its velocity. Sketch the velocity vector for a displacement in each quadrant.

$$Ans.\quad u = -\frac{2\pi \, \overline{\Delta y}^3}{3(\overline{\Delta x}^2 + \overline{\Delta y}^2)}; v = -\frac{2\pi \, \overline{\Delta x}^3}{3(\overline{\Delta x}^2 + \overline{\Delta y}^2)}.$$

8. Find the velocity at $(0,2b)$ due to a double row of vortices of strength κ located at $(0, \pm b)$, $(\pm a, \pm b)$, $(\pm 2a, \pm b)$, $(\pm 3a, \pm b)$, ...

$$Ans.\quad u = \frac{\kappa}{2a}\left(\coth\frac{3\pi b}{a} + \coth\frac{\pi b}{a}\right); v = 0.$$

9. Find the velocity at the origin due to constant vorticity $\zeta = 2$ sec^{-1} in the unit square with sides parallel to the coordinate axes and center at $(\tfrac{1}{2},\tfrac{1}{2})$.

$Ans.$ $u = 0.36$ ft per sec; $v = -0.36$ ft per sec.

10. Find the velocity at the origin due to the semicircular area of vorticity above the x-axis with center at the origin. The radius of the semicircle is 10 ft, and the vorticity is given by $\zeta = 2r$ sec^{-1}. $Ans.$ $u = 31.8$ ft per sec; $v = 0$.

CHAPTER X

EQUATIONS FOR VISCOUS FLOW

In the preceding chapters an ideal fluid has, in general, been assumed. These assumptions permit the solution of many flow cases that apply closely to the motion of real fluids with low viscosity. The effects of viscosity in these cases are limited to the immediate neighborhood of boundaries. The solutions are usually valid for the first stages of the flow, *i.e.*, as it is initiated and before the boundary layer has developed. Ideal fluid theory gives no information as to the drag on an object moving relative to the fluid.

All evidence indicates[1] that an actual fluid in contact with a solid boundary has no motion relative to the boundary. Due to the adhesion of the fluid to the solid, fluid is retarded near the boundary, relative to the boundary. This zone of retarded fluid is called the *boundary layer*.

The equations of motion, when viscosity is taken into account, are derived in this chapter. The resulting simultaneous nonlinear partial differential equations are so complex that their solution can be effected for only the most simple flow cases. Before undertaking their derivation, the concept of stress is reviewed.

ANALYTICAL STATICS OF A THREE-DIMENSIONAL CONTINUUM

The concepts and relationships of shear and normal stresses are essential in the development of the equations of motion. Following the lecture notes of Prof. M. Sadowsky in his graduate course in analytical mechanics at Illinois Institute of Technology, the relationships are developed for static bodies (continua). By use of D'Alembert's principle of dynamic equilibrium the same relationships are found to apply to moving bodies. No limiting assumptions are made as to the nature of the continuum, whether solid, liquid, plastic, or gas, other than those given in Sec. 2.

107. Notion of Inner Stress. When a body is acted upon by external forces and an internal cut is made in the body, a gap may occur, as shown in Fig. 117. The cut has removed the action of internal forces that were holding the faces together. Those internal forces in a body which

---

[1] S. Goldstein (ed.), "Modern Developments in Fluid Dynamics," pp. 676–680, Oxford University Press, New York, 1938.

prevent any gap from occurring and which prevent motion of one face relative to the other are called *stress forces*.   At any face they are equal and opposite, as required by Newton's third law.   Inner stress forces arise from the action of external forces.   A stress is an internal intensity

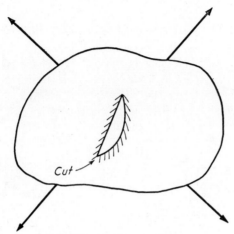

FIG. 117.—Cut in a continuum.

of force expressed as force per unit area.   If the resulting stress is normal to the cut, it is called a *normal stress;* if it is tangent to the cut, it is a *shear stress*.   Any general stress may be decomposed into normal and shear stress components.

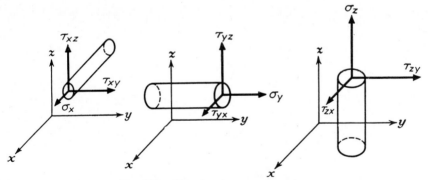

FIG. 118.—The nine stress components.

108. Stress Components in a Cartesian Coordinate System.   There are nine stress components, which are evident when three sections through the continuum are taken at a point, each section normal to a coordinate axis of a cartesian coordinate system.   These are shown in Fig. 118, in which the stress components parallel to the axes are taken for the three planes. Six of the nine stresses are shown to be independent. The normal stresses are indicated by σ, considered positive (tension) when in the direction of the outer normal and carrying a subscript to indicate its direction. A negative normal stress (compression) is in the direction of the inner normal. The tangential stresses

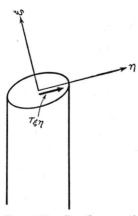

(shear stresses) are indicated by τ. The sign of a shear stress has no physical significance; it is a mathematical formality. The first subscript indicates the direction of the normal to the plane over which the stress acts, and the second subscript indicates the direction of the stress, as shown in Fig. 119. The sign convention is as follows: Referring to Fig. 120, $\tau_{\eta\xi}$ is positive if acting in the direction of negative ξ-axis on a face preceding the direction of positive η-axis. All shear stresses shown in Fig. 120 are positive.

Fig. 119.—Significance of shear stress subscripts.

A small prismatic body cut out of the continuum has the nine stresses shown in Fig. 121. Taking moments about an axis through the center of the body parallel to the z-axis shows that $\tau_{xy} = \tau_{yx}$, as follows:

$$\Sigma M = 0$$

The only stresses acting to cause moments are τ_{xy} and τ_{yx}. Letting the edges of the prism be δx, δy, δz, respectively, parallel to the xyz-axes, the moment equation becomes

$$-\tau_{yx}\,\delta x\,\delta y\,\delta z + \tau_{xy}\,\delta y\,\delta z\,\delta x = 0$$

hence,

$$\tau_{xy} = \tau_{yx} \qquad (1)$$

and similarly,

$$\tau_{yz} = \tau_{zy}, \qquad \tau_{zx} = \tau_{xz} \qquad (2)$$

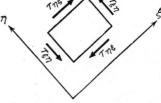

Fig. 120.—Sign convention for shear stresses. Those shown are positive.

For a moving body the summation of moments is equated to the product of the moment of inertia of the body and its angular acceleration. The moment of inertia term is of higher order of smallness than the moment term, so that it drops out in the limit as δx, δy, δz approach zero, leaving Eqs. (1) and (2) valid for moving bodies.

It must be proved that the nine stress components completely define the state of stress at a point.   Using a tetrahedron (Fig. 122), since it is the solid with least number of faces, the stresses on an arbitrarily inclined

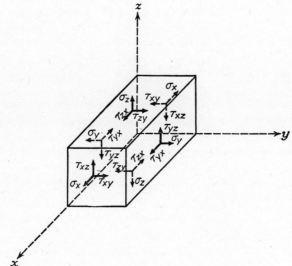

Fig. 121.—Prismatic body showing stress components.

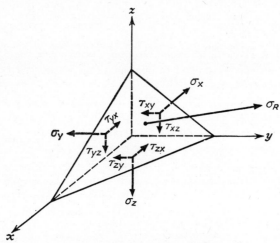

Fig. 122.—Stresses on tetrahedron.

face are expressed in terms of the stress components.   Let l, m, n be the direction cosines of the normal to the inclined face; let the area of the inclined face be A; and let the xyz-components of the stress σ_R on the inclined face be σ_{Rx}, σ_{Ry}, σ_{Rz}, respectively. Applying the equations of equilibrium

$$\Sigma F = 0$$

in the directions of the coordinate axes yields (from Fig. 122)

$$\left. \begin{array}{l} \sigma_{Rx} = \sigma_x l + \tau_{yx} m + \tau_{zx} n \\ \sigma_{Ry} = \tau_{xy} l + \sigma_y m + \tau_{zy} n \\ \sigma_{Rz} = \tau_{xz} l + \tau_{yz} m + \sigma_z n \end{array} \right\} \qquad (3)$$

Since the stresses on any arbitrarily inclined face are completely given in terms of the nine stress components and the direction cosines of the normal to the inclined face, it has been proved that they completely define the state of stress at a point.

Due to Eqs. (1) and (2), there are then six independent stress components that are required to determine completely the state of stress at a point. The stress components may be conveniently written as a matrix

$$\begin{array}{ccc} \sigma_x & \tau_{xy} & \tau_{xz} \\ \tau_{yx} & \sigma_y & \tau_{yz} \\ \tau_{zx} & \tau_{zy} & \sigma_z \end{array}$$

which also completely defines the state of stress at a point.

109. Principal Stresses. In this section it is shown that for any general state of stress, there are, at every point, at least three planes over which the shear stresses vanish. The resulting stresses on these planes are normal stresses and are called *principal stresses.* Axes through the point coincident with the principal stress directions are called *principal axes,* and the planes over which they act *principal planes.* Thus, by definition, the stresses are perpendicular to the faces over which they act, and likewise any stress that acts normal to a face in the absence of other components is referred to as a principal stress. A body subjected to principal stresses is easily visualized, since the forces on the surface are normal to the faces. The converse case is examined here, *viz.,* with a continuum subjected to the most general system of surface stresses, to show that at every point a special position of a rectangular element around the point may be found such that only normal stresses act on its faces.

First, the relation between the relative size of the principal stresses and the inclinations of the principal axes are determined. Assume that two or more principal planes exist, and let θ be the angle between them, as in Fig. 123.   Taking moments about O for the element shown, where c is the dimension normal to the plane of the drawing,

$$\Sigma M = 0$$

i.e.,

$$-ac\sigma_1 b \cos\theta + bc\sigma_2 a \cos\theta = 0$$

Simplifying,

$$(\sigma_2 - \sigma_1) \cos\theta = 0.$$

This relationship may be satisfied in two ways:

1. $\sigma_2 \neq \sigma_1$, then $\cos\theta = 0$, $\theta = 90$ deg.   This is the general case in which all principal axes are perpendicular at a point, which limits their number to three.

2. $\sigma_2 = \sigma_1$, θ arbitrary.   This is the ideal fluid flow case in which the stress is the same in all directions at a point.

To prove the existence of three principal stresses, their existence is assumed and the necessity of existence proved.   Referring to Fig. 122, assume that $\sigma_R = \sigma$ is a principal stress. σ must be perpendicular to the face, *i.e.*, along the normal whose direction cosines are l, m, n.   The components of σ in the xyz-directions then are

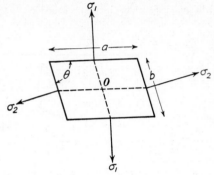

FIG. 123.—Principal stresses on an element.

$$\left.\begin{array}{l} \sigma l = \sigma_x l + \tau_{xy} m + \tau_{xz} n \\ \sigma m = \tau_{xy} l + \sigma_y m + \tau_{yz} n \\ \sigma n = \tau_{xz} l + \tau_{yz} m + \sigma_z n \end{array}\right\} \qquad (4)$$

Rearranging,

$$\left.\begin{array}{l} (\sigma_x - \sigma)l + \tau_{xy} m + \tau_{xz} n = 0 \\ \tau_{xy} l + (\sigma_y - \sigma)m + \tau_{yz} n = 0 \\ \tau_{xz} l + \tau_{yz} m + (\sigma_z - \sigma)n = 0 \end{array}\right\} \qquad (5)$$

Thus three homogeneous linear equations in l, m, n are obtained.   The trivial solution is impossible, $l = m = n = 0$, since

$$l^2 + m^2 + n^2 = 1$$

Using the determinate method of solution, it is necessary that the denominator determinate be zero; otherwise there could be no solution.   Hence,

$$D = \begin{vmatrix} (\sigma_x - \sigma) & \tau_{xy} & \tau_{xz} \\ \tau_{xy} & (\sigma_y - \sigma) & \tau_{yz} \\ \tau_{xz} & \tau_{yz} & (\sigma_z - \sigma) \end{vmatrix} = 0 \qquad (6)$$

Expansion of Eq. (6) leads to a cubic equation in σ:

$$D = F(\sigma) = 0$$

If it can be shown that Eq. (6) has three real, distinct roots, it is evident that Eqs. (5) yield three sets of direction cosines corresponding to the three principal stresses. Inspection of Eq. (6) shows that the highest degree term of the expansion is $-\sigma^3$. Thus, if $\sigma \to +\infty$, $F(\sigma)$ is negative; and if $\sigma \to -\infty$, $F(\sigma)$ is positive, proving at least one real root exists.

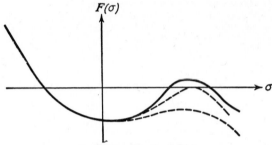

Fig. 124.—Graph of $F(\sigma)$.

The remaining two may be distinct real, coincident real, or conjugate complex, as illustrated in Fig. 124. Knowing that σ_1 (the real root) exists, the arbitrary xyz-system may be chosen such that the z-axis coincides with σ_1, so that $\tau_{zx} = \tau_{zy} = 0$, $\sigma_z = \sigma_1$. For this system the denominator determinate of Eq. (5) reduces to

$$D_1 = \begin{vmatrix} (\sigma_x - \sigma) & \tau_{xy} & 0 \\ \tau_{xy} & (\sigma_y - \sigma) & 0 \\ 0 & 0 & (\sigma_1 - \sigma) \end{vmatrix} = 0 \tag{7}$$

All the stress components have necessarily changed, but not the principal stresses, which are independent of the choice of reference axes and depend entirely on the state of stress. Equation (7) leads to a quadratic in σ, viz.,

$$\sigma^2 - \sigma(\sigma_x + \sigma_y) + \sigma_x\sigma_y - \tau_{xy}^2 = 0$$

Solving,

$$\sigma = \frac{\sigma_x + \sigma_y \pm \sqrt{(\sigma_x - \sigma_y)^2 + 4\tau_{xy}^2}}{2}$$

Since the discriminant of the quadratic (portion under the radical) is positive, sum of two squares, $D_1 = 0$ has two real, distinct roots.[1] Hence, $D = 0$ has three real, distinct roots.

---

[1] A special case arises when the discriminant is zero; *i.e.*, $\tau_{xy} = 0$, and $\sigma_x = \sigma_y$. This is the ideal fluid case, where the only stresses are equal principal stresses.

There must exist, in general, three distinct principal stresses, corresponding to three mutually perpendicular axes in space, given by substituting in turn σ_1, σ_2, σ_3 in Eqs. (5). Selecting the principal axes as axes of reference, the general stress matrix reduces to

$$\begin{matrix} \sigma_1 & 0 & 0 \\ 0 & \sigma_2 & 0 \\ 0 & 0 & \sigma_3 \end{matrix}$$

The invariants in the transformation of a state of stress are obtained from the expansion of Eq. (6). The state of stress at a point may be expressed in terms of an arbitrary coordinate system or referred to the principal axes, 1, 2, 3. Expanding Eq. (6),

$$\sigma^3 - \underbrace{(\sigma_x + \sigma_y + \sigma_z)}_{I_1}\sigma^2 + \underbrace{(\sigma_x\sigma_y + \sigma_y\sigma_z + \sigma_z\sigma_x - \tau_{xy}^2 - \tau_{yz}^2 - \tau_{zz}^2)}_{I_2}\sigma$$

$$- \underbrace{\begin{vmatrix} \sigma_x & \tau_{xy} & \tau_{xz} \\ \tau_{xy} & \sigma_y & \tau_{yz} \\ \tau_{xz} & \tau_{yz} & \sigma_z \end{vmatrix}}_{I_3} = 0 \quad (8)$$

This cubic determines the values of σ_1, σ_2, σ_3. Since these values are independent of the choice of axes and inherent in the state of stress, the quantities I_1, I_2, I_3 in Eq. (8) must be *invariant under any rotation of the coordinate system.* Hence,

$$I_1 = \sigma_x + \sigma_y + \sigma_z = \sigma_1 + \sigma_2 + \sigma_3 \quad (9)$$

$$I_2 = \sigma_x\sigma_y + \sigma_y\sigma_z + \sigma_z\sigma_x - \tau_{xy}^2 - \tau_{yz}^2 - \tau_{zz}^2$$
$$= \sigma_1\sigma_2 + \sigma_2\sigma_3 + \sigma_3\sigma_1 \quad (10)$$

$$I_3 = \begin{vmatrix} \sigma_x & \tau_{xy} & \tau_{xz} \\ \tau_{xy} & \sigma_y & \tau_{yz} \\ \tau_{xz} & \tau_{yz} & \sigma_z \end{vmatrix} = \sigma_1\sigma_2\sigma_3 \quad (11)$$

Thus, the equation for determination of the principal stresses, σ_1, σ_2, σ_3 may be written as

$$\sigma^3 - I_1\sigma^2 + I_2\sigma - I_3 = 0$$

EQUATIONS OF MOTION

In order to obtain the equations of motion for viscous fluids, the relationships between the stresses and velocity gradients must be established. An assumption must be made in regard to these relationships. The equations are first developed in terms of the stresses at a point, then the viscosity definition is extended to the general case, next the normal stresses are expressed in terms of the viscosity and velocity gradients, and finally, by substitution, the Navier-Stokes equations are obtained.

110. Equations of Motion in Terms of Stress Components. The effect of viscosity is to cause shear stresses in the fluid. Referring to Fig. 125, the shear stresses on a small rectangular parallelepiped may be included in the equations of motion as follows:

Let the stresses at the center of the parallelepiped be τ_{xy}, τ_{yz}, τ_{zx}, σ_x,

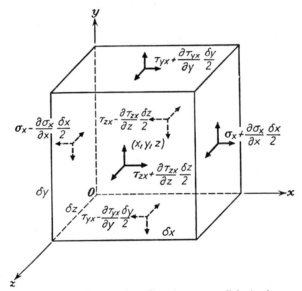

FIG. 125.—Stresses in x-direction on parallelepiped.

σ_y, σ_z. At the face normal to the y-axis and most distant from the origin, the shear stress in the x-direction is

$$\tau_{yx} + \frac{\partial \tau_{yx}}{\partial y}\frac{\delta y}{2}$$

and the shear force on the face is

$$\left(\tau_{yx} + \frac{\partial \tau_{yx}}{\partial y}\frac{\delta y}{2}\right)\delta x\ \delta z$$

Similarly, the shear force on the opposite face in the x-direction is

$$-\left(\tau_{yx} - \frac{\partial \tau_{yx}}{\partial y}\frac{\delta y}{2}\right)\delta x\ \delta z$$

Likewise, the shear forces in the x-direction on the faces normal to the z-axis are

$$\left(\tau_{zx} + \frac{\partial \tau_{zx}}{\partial z}\frac{\delta z}{2}\right)\delta x\ \delta y$$

and

$$-\left(\tau_{zx} - \frac{\partial \tau_{zx}}{\partial z}\frac{\delta z}{2}\right)\delta x\ \delta y$$

Adding to these forces the other forces in the x-direction and equating to the product of the mass and the x-component of acceleration, as in Sec. 6,

$$X\rho\ \delta x\ \delta y\ \delta z + \frac{\partial \sigma_x}{\partial x}\ \delta x\ \delta y\ \delta z + \frac{\partial \tau_{yx}}{\partial y}\ \delta x\ \delta y\ \delta z + \frac{\partial \tau_{zx}}{\partial z}\ \delta x\ \delta y\ \delta z = \rho\ \delta x\ \delta y\ \delta z\ a_x.$$

Dividing through by the mass of the particle and taking the limit as the parallelepiped shrinks to a point, the first of the three equations of motion are obtained:

$$\left. \begin{aligned} X + \frac{1}{\rho}\frac{\partial \sigma_x}{\partial x} + \frac{1}{\rho}\frac{\partial \tau_{yx}}{\partial y} + \frac{1}{\rho}\frac{\partial \tau_{zx}}{\partial z} = a_x \\ Y + \frac{1}{\rho}\frac{\partial \tau_{xy}}{\partial x} + \frac{1}{\rho}\frac{\partial \sigma_y}{\partial y} + \frac{1}{\rho}\frac{\partial \tau_{zy}}{\partial z} = a_y \\ Z + \frac{1}{\rho}\frac{\partial \tau_{xz}}{\partial x} + \frac{1}{\rho}\frac{\partial \tau_{yz}}{\partial y} + \frac{1}{\rho}\frac{\partial \sigma_z}{\partial z} = a_z \end{aligned} \right\} \qquad (12)$$

The shear and normal stresses must now be expressed in terms of velocity gradients and viscosity further to reduce these equations.

111. Viscosity.   In Sec. 1 Newton's law of viscosity for one-dimensional flow was stated.   Viscosity is that property of a fluid by virtue of which it offers resistance to shear.   In the one-dimensional case it was shown that the shear stress is a linear function of the time rate of angular deformation.

It is assumed here that in general flow the shear stress components are proportional to the corresponding time rates of angular deformation. Referring to Fig. 126 the time rate of angular deformation is determined from the angular velocities of the two adjacent sides of the element. The angular velocity of the linear element δy is $-\dfrac{\partial u}{\partial y}$ and of δx is $\dfrac{\partial v}{\partial x}$ as in Sec. 9, considering the counterclockwise direction positive.   The rate of angular deformation is then

$$\frac{\partial v}{\partial x} + \frac{\partial u}{\partial y}$$

Extending Newton's law of viscosity to this general case,

$$\left.\begin{aligned} \tau_{xy} = \tau_{yx} &= \mu\left(\frac{\partial v}{\partial x} + \frac{\partial u}{\partial y}\right) \\ \tau_{yz} = \tau_{zy} &= \mu\left(\frac{\partial w}{\partial y} + \frac{\partial v}{\partial z}\right) \\ \tau_{zx} = \tau_{xz} &= \mu\left(\frac{\partial u}{\partial z} + \frac{\partial w}{\partial x}\right) \end{aligned}\right\} \tag{13}$$

The expressions for normal stress in terms of the velocity gradients and viscosity are developed in the following section.

112. Relation between Normal Stresses and Velocity Gradients. Referring to Eqs. (13) and Fig. 126, another selection of axes could have been made such that $\tau_{xy} = \tau_{yz} = \tau_{zx} = 0$. These are the principal axes, discussed in Sec. 109. For these principal axes, denoted by x', y', z' and with velocity components u', v', w',

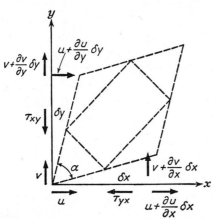

Fig. 126.—Angular deformation in two-dimensional flow.

$$\frac{\partial v'}{\partial x'} = -\frac{\partial u'}{\partial y'}, \qquad \frac{\partial w'}{\partial y'} = -\frac{\partial v'}{\partial z'},$$

$$\frac{\partial u'}{\partial z'} = -\frac{\partial w'}{\partial x'} \tag{14}$$

The orientation of principal axes with respect to the arbitrary xyz-axes are given by the table of direction cosines. For example, the y'-axis has the direction cosines l_2, m_2, n_2 referred to the xyz-system.

	x	y	z
x'	l_1	m_1	n_1
y'	l_2	m_2	n_2
z'	l_3	m_3	n_3

Along the principal axes there is only a linear deformation of an element. Taking a small element with sides parallel to the principal axes, as in Fig. 127, Eqs. (14) show that the time rate of angular deformation is zero. The element may translate, rotate, and the sides expand or contract, but the angles of the element remain right angles during the infinitesimal time the sides are in the principal axes. On the other hand, assuming the principal axes to rotate with the element, the extension of an element along a principal axis depends upon that coordinate only; therefore, u' is a function of x' only, v' is a function of y' only, and w' is a function of z' only; hence, each of the terms in Eqs. (14) is equal to zero, while in general $u = u(x,y,z)$, $v = v(x,y,z)$, and $w = w(x,y,z)$.

The velocity components and independent variables in the two systems are related as follows:

$$
\begin{aligned}
u &= u'l_1 + v'l_2 + w'l_3 \\
v &= u'm_1 + v'm_2 + w'm_3 \\
w &= u'n_1 + v'n_2 + w'n_3 \\
x' &= l_1 x + m_1 y + n_1 z \\
y' &= l_2 x + m_2 y + n_2 z \\
z' &= l_3 x + m_3 y + n_3 z
\end{aligned}
$$

since in each case the component in a given direction is given by the components in that direction of any three mutually perpendicular components. The velocity gradients in the arbitrary system are to be expressed in terms of the principal axes and their velocity components as a means of simplifying the relationships and the assumption. The expression $\dfrac{\partial u}{\partial x}$ may be written

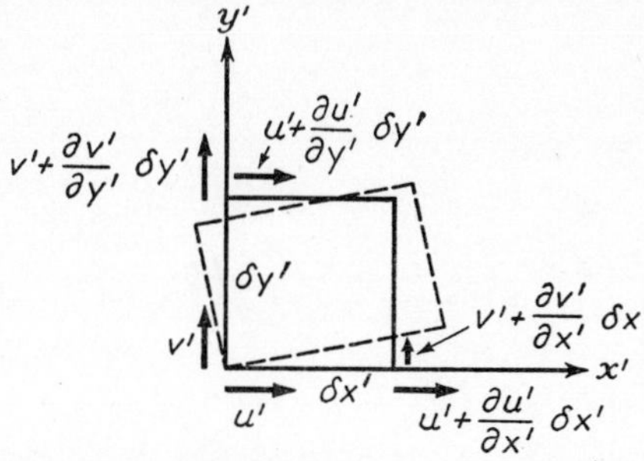

Fig. 127.—Motion of element with sides parallel to principal axes.

$$
\frac{\partial u}{\partial x} = \frac{\partial u}{\partial x'}\frac{\partial x'}{\partial x} + \frac{\partial u}{\partial y'}\frac{\partial y'}{\partial x} + \frac{\partial u}{\partial z'}\frac{\partial z'}{\partial x}
$$

since x is a function of x', y', z'. Using the relation between coordinates,

$$
\frac{\partial x'}{\partial x} = l_1, \qquad \frac{\partial y'}{\partial x} = l_2, \qquad \frac{\partial z'}{\partial x} = l_3
$$

Hence,

$$
\frac{\partial u}{\partial x} = \left(\frac{l_1 \partial}{\partial x'} + \frac{l_2 \partial}{\partial y'} + \frac{l_3 \partial}{\partial z'} \right)(u'l_1 + v'l_2 + w'l_3)
$$

and

$$
\left.
\begin{aligned}
\frac{\partial u}{\partial x} &= l_1{}^2 \frac{\partial u'}{\partial x'} + l_2{}^2 \frac{\partial v'}{\partial y'} + l_3{}^2 \frac{\partial w'}{\partial z'} \\
\frac{\partial v}{\partial y} &= m_1{}^2 \frac{\partial u'}{\partial x'} + m_2{}^2 \frac{\partial v'}{\partial y'} + m_3{}^2 \frac{\partial w'}{\partial z'} \\
\frac{\partial w}{\partial z} &= n_1{}^2 \frac{\partial u'}{\partial x'} + n_2{}^2 \frac{\partial v'}{\partial y'} + n_3{}^2 \frac{\partial w'}{\partial z'}
\end{aligned}
\right\} \qquad (15)
$$

using Eqs. (14) or, for rotating $x'y'z'$-axes, u' is a function of x' only, etc.

The last two equations are obtained in a similar manner.

Adding Eqs. (15),

$$\frac{\partial u}{\partial x} + \frac{\partial v}{\partial y} + \frac{\partial w}{\partial z} = \frac{\partial u'}{\partial x'} + \frac{\partial v'}{\partial y'} + \frac{\partial w'}{\partial z'} = \text{div } q \tag{16}$$

which shows that the expansion of fluid at a point is independent of the orientation of axes.

Furthermore,

$$\frac{\partial u}{\partial y} = \left(\frac{m_1 \partial}{\partial x'} + \frac{m_2 \partial}{\partial y'} + \frac{m_3 \partial}{\partial z'} \right) (u'l_1 + v'l_2 + w'l_3)$$

and

$$\frac{\partial v}{\partial x} = \left(\frac{l_1 \partial}{\partial x'} + \frac{l_2 \partial}{\partial y'} + \frac{l_3 \partial}{\partial z'} \right) (u'm_1 + v'm_2 + w'm_3)$$

Adding the two expressions,

$$\left. \begin{aligned} \frac{\partial v}{\partial x} + \frac{\partial u}{\partial y} &= 2 \left(l_1 m_1 \frac{\partial u'}{\partial x'} + l_2 m_2 \frac{\partial v'}{\partial y'} + l_3 m_3 \frac{\partial w'}{\partial z'} \right) \\ \frac{\partial w}{\partial y} + \frac{\partial v}{\partial z} &= 2 \left(m_1 n_1 \frac{\partial u'}{\partial x'} + m_2 n_2 \frac{\partial v'}{\partial y'} + m_3 n_3 \frac{\partial w'}{\partial z'} \right) \\ \frac{\partial u}{\partial z} + \frac{\partial w}{\partial x} &= 2 \left(n_1 l_1 \frac{\partial u'}{\partial x'} + n_2 l_2 \frac{\partial v'}{\partial y'} + n_3 l_3 \frac{\partial w'}{\partial z'} \right) \end{aligned} \right\} \tag{17}$$

where the last two expressions are obtained in a similar manner. The shear stresses may now be expressed in terms of the velocity gradients referred to principal axes.

The shear stresses are next expressed in terms of the principal stresses $\sigma_{x'}$, $\sigma_{y'}$, $\sigma_{z'}$. Writing the equations of motion for the tetrahedron of Fig. 128, which has three faces in the principal planes over which no shear stresses act and the inclined face normal to the x-axis, and remembering that the terms containing the mass are of higher order in the limit as the inclined plane approaches the origin,

$$\sigma_x(ABC) = \sigma_{x'}(OBC)l_1 + \sigma_{y'}(OAC)l_2 + \sigma_{z'}(OAB)l_3$$

since

$$(OBC) = l_1(ABC), \qquad (OAC) = l_2(ABC), \qquad (OAB) = l_3(ABC)$$

$$\sigma_x = \sigma_{x'}l_1{}^2 + \sigma_{y'}l_2{}^2 + \sigma_{z'}l_3{}^2 \tag{18a}$$

Similarly, by taking the inclined face normal to the y- and the z-axis, respectively,

$$\sigma_y = \sigma_{x'}m_1{}^2 + \sigma_{y'}m_2{}^2 + \sigma_{z'}m_3{}^2 \tag{18b}$$

$$\sigma_z = \sigma_{x'}n_1{}^2 + \sigma_{y'}n_2{}^2 + \sigma_{z'}n_3{}^2 \tag{18c}$$

Adding Eqs. (18),

$$\sigma_x + \sigma_y + \sigma_z = \sigma_{x'} + \sigma_{y'} + \sigma_{z'} = -3p \qquad (19)$$

which is Eq. (9).   The sum of the normal stresses at a point is a constant. The average pressure intensity at a point is p, as defined by the equation. In nonviscous flow the normal stresses are the same in all directions.

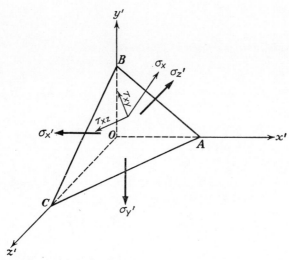

Fig. 128.—Tetrahedron with inclined face normal to x-axis and other faces in principal planes.

Referring again to Fig. 128, to express the shear stresses in terms of the principal stresses,

$$\tau_{xy}(ABC) = \sigma_{x'}(OBC)m_1 + \sigma_{y'}(OAC)m_2 + \sigma_{z'}(OAB)m_3$$

and reducing as before,

$$\tau_{xy} = \sigma_{x'}l_1m_1 + \sigma_{y'}l_2m_2 + \sigma_{z'}l_3m_3 \qquad (20)$$

The shear stress τ_{xy} can now be expressed two ways

$$
\begin{aligned}
\tau_{xy} &= \mu\left(\frac{\partial u}{\partial y} + \frac{\partial v}{\partial x}\right) \\
&= 2\mu\left(l_1m_1\frac{\partial u'}{\partial x'} + l_2m_2\frac{\partial v'}{\partial y'} + l_3m_3\frac{\partial w'}{\partial z'}\right) \\
&= \sigma_{x'}l_1m_1 + \sigma_{y'}l_2m_2 + \sigma_{z'}l_3m_3
\end{aligned}
\qquad (21)
$$

The normal stresses must be expressed in terms of the viscosity and velocity gradients in some manner that is compatible with Eq. (21). The most general linear relationship is

$$\left.\begin{aligned} \sigma_{x'} &= -p + a\left(\frac{\partial u'}{\partial x'} + \frac{\partial v'}{\partial y'} + \frac{\partial w'}{\partial z'}\right) + b\frac{\partial u'}{\partial x'} \\ \sigma_{y'} &= -p + a\left(\frac{\partial u'}{\partial x'} + \frac{\partial v'}{\partial y'} + \frac{\partial w'}{\partial z'}\right) + b\frac{\partial v'}{\partial y'} \\ \sigma_{z'} &= -p + a\left(\frac{\partial u'}{\partial x'} + \frac{\partial v'}{\partial y'} + \frac{\partial w'}{\partial z'}\right) + b\frac{\partial w'}{\partial z'} \end{aligned}\right\} \tag{22}$$

Adding Eqs. (22) and comparing with Eq. (19), it is seen that

$$3a = -b$$

Substituting Eqs. (22) into Eq. (21), remembering that

$$l_1 m_1 + l_2 m_2 + l_3 m_3 = 0$$

since the two lines represented by the direction cosines are perpendicular,

$$b = 2\mu$$

Substituting the values of a and b into Eqs. (22) and then expressing the velocity gradients in terms of the arbitrary xyz-system, using Eqs. (15), (16), and (18),

$$\left.\begin{aligned} \sigma_x &= -p - \frac{2}{3}\mu\left(\frac{\partial u}{\partial x} + \frac{\partial v}{\partial y} + \frac{\partial w}{\partial z}\right) + 2\mu\frac{\partial u}{\partial x} \\ \sigma_y &= -p - \frac{2}{3}\mu\left(\frac{\partial u}{\partial x} + \frac{\partial v}{\partial y} + \frac{\partial w}{\partial z}\right) + 2\mu\frac{\partial v}{\partial y} \\ \sigma_z &= -p - \frac{2}{3}\mu\left(\frac{\partial u}{\partial x} + \frac{\partial v}{\partial y} + \frac{\partial w}{\partial z}\right) + 2\mu\frac{\partial w}{\partial z} \end{aligned}\right\} \tag{23}$$

113. Navier-Stokes Equations. Using the expressions for shear and normal stresses in terms of the viscosity and the velocity gradients, as given by Eqs. (13) and (23), substitution into Eqs. (12) gives the Navier-Stokes equations

$$\left.\begin{aligned} X - \frac{1}{\rho}\frac{\partial p}{\partial x} + \frac{\nu}{3}\frac{\partial}{\partial x}\left(\frac{\partial u}{\partial x} + \frac{\partial v}{\partial y} + \frac{\partial w}{\partial z}\right) + \nu\,\nabla^2 u &= \frac{Du}{Dt} \\ Y - \frac{1}{\rho}\frac{\partial p}{\partial y} + \frac{\nu}{3}\frac{\partial}{\partial y}\left(\frac{\partial u}{\partial x} + \frac{\partial v}{\partial y} + \frac{\partial w}{\partial z}\right) + \nu\,\nabla^2 v &= \frac{Dv}{Dt} \\ Z - \frac{1}{\rho}\frac{\partial p}{\partial z} + \frac{\nu}{3}\frac{\partial}{\partial z}\left(\frac{\partial u}{\partial x} + \frac{\partial v}{\partial y} + \frac{\partial w}{\partial z}\right) + \nu\,\nabla^2 w &= \frac{Dw}{Dt} \end{aligned}\right\} \tag{24}$$

in which $\nu = \mu/\rho$ is the kinematic viscosity and

$$\nabla^2 = \frac{\partial^2}{\partial x^2} + \frac{\partial^2}{\partial y^2} + \frac{\partial^2}{\partial z^2}$$

These equations are the most general ones for fluid flow. By letting $\nu = 0$ they reduce to the Euler equations. For incompressible fluids the term containing the divergence drops out, leaving

$$\left.\begin{array}{l} X - \dfrac{1}{\rho}\dfrac{\partial p}{\partial x} + \nu\,\nabla^2 u = \dfrac{Du}{Dt} \\[2mm] Y - \dfrac{1}{\rho}\dfrac{\partial p}{\partial y} + \nu\,\nabla^2 v = \dfrac{Dv}{Dt} \\[2mm] Z - \dfrac{1}{\rho}\dfrac{\partial p}{\partial z} + \nu\,\nabla^2 w = \dfrac{Dw}{Dt} \end{array}\right\} \tag{25}$$

In deriving the equations the assumption was made that the stresses are linear functions of the velocity gradients. Due to the complexity of the equations, no solutions that would confirm the validity of the assumption have been effected for a general case. For those solutions which are known, obtained by neglecting terms in the equations, the results are confirmed by experiment. The equations were first derived by Navier and Poisson by an entirely different method in 1822 and 1829. They were derived in a manner similar to the above method by Saint-Venant and Stokes in 1843 and 1845.

For a specific problem the solution must satisfy not only the Navier-Stokes equations but also the continuity equation and the boundary conditions for the particular problem.

114. Boundary Conditions. It is generally accepted that the fluid in contact with a solid boundary moves with the boundary or that the velocity of fluid at the boundary relative to the boundary is zero. This hypothesis has been borne out by experiment where it is possible to make reductions in the Navier-Stokes equations that permit their solution.

When two fluids are flowing, a dynamical boundary condition arises at the interface. Applying the equation of motion to a thin layer of fluid enclosing a small portion of the interface shows that the terms containing the mass are of higher order than the surface intensities and that the stresses must be continuous through the interface.

In general, the boundary conditions at the solid surfaces give rise to rotational flow. This precludes the simplification that results from the use of a velocity potential. Examples of simplified solutions of the equations of viscous flow are given in the following chapter.

CHAPTER XI

EXAMPLES OF VISCOUS FLOW

Since no general solutions of the Navier-Stokes equations are known, the examples of this chapter in each case are obtained by neglecting certain terms in the differential equations. Similitude relationships are first discussed, followed by flow between parallel boundaries, the lubrication problem, flow through circular tubes, percolation, and flow around a sphere at low Reynolds numbers.

115. Similitude Relationships: Reynolds Number. Due to the complexity of the viscous flow equations, it has been necessary to resort to experimental means for the solution of many fluid problems. Even though the Navier-Stokes equations have not been solved, much can be learned from them through considerations of similarity. For two flow cases to be dynamically similar the following criteria must be met:

1. The geometrical boundaries must be similar.
2. The boundary conditions must be the same.
3. The streamlines, or flow patterns, must be geometrically similar, or the dynamic pressures at corresponding points must bear a fixed ratio to each other.

In applying the Navier-Stokes equations to the two flow cases, the alteration of units of length, time, and pressure should transform one equation into the other for complete similarity. Letting subscript 1 refer to one flow and subscript 2 to the other, the ratios of length, time, and pressure scales may be represented by the dimensionless quantities ξ, τ, π, respectively; thus

$$\frac{l_1}{l_2} = \xi, \qquad \frac{t_1}{t_2} = \tau, \qquad \frac{p_1}{p_2} = \pi \tag{1}$$

The length ratio applies to the lengths between corresponding points, while the time ratio applies to the time for a fluid particle to travel corresponding distances. The pressures are considered to be dynamic pressures only, since the effect of gravity and the static portion of the total pressure terms compensate.

Omitting the extraneous force term from the first of the Navier-Stokes equations for incompressible flow [Eqs. (25), Chap. X],

$$-\frac{1}{\rho}\frac{\partial p}{\partial x} + \frac{\mu}{\rho}\left(\frac{\partial^2 u}{\partial x^2} + \frac{\partial^2 u}{\partial y^2} + \frac{\partial^2 u}{\partial z^2}\right) = u\frac{\partial u}{\partial x} + v\frac{\partial u}{\partial y} + w\frac{\partial u}{\partial z} + \frac{\partial u}{\partial t} \tag{2}$$

The development is limited to fluids of constant density and viscosity, with their ratios taken as

$$\mu' = \frac{\mu_1}{\mu_2}, \qquad \rho' = \frac{\rho_1}{\rho_2}$$

The velocity term contains the units length over time; hence, the velocity ratio is given by the length ratio divided by the time ratio

$$\frac{\xi}{\tau} = \eta \sim \frac{u_1}{u_2} \sim \frac{v_1}{v_2} \sim \frac{w_1}{w_2}$$

Acceleration is velocity divided by time and can be represented by

$$\frac{\eta}{\tau} = \frac{\xi}{\tau^2} = \frac{\eta^2}{\xi}$$

The following similitude relationships are to be substituted into Eq. (2) when written with subscripts 1:

$$x_1 = \xi x_2, \ldots, \qquad u_1 = \eta u_2, \ldots$$
$$p_1 = \pi p_2, \qquad \rho_1 = \rho' \rho_2, \qquad \mu_1 = \mu' \mu_2$$

This yields

$$-\frac{\pi}{\rho'\xi}\frac{1}{\rho_2}\frac{\partial p_2}{\partial x_2} + \frac{\mu'\eta}{\rho'\xi^2}\frac{\mu_2}{\rho_2}\left(\frac{\partial^2 u_2}{\partial x_2{}^2} + \frac{\partial^2 u_2}{\partial y_2{}^2} + \frac{\partial^2 u_2}{\partial z_2{}^2}\right)$$
$$= \frac{\eta^2}{\xi}\left(u_2\frac{\partial u_2}{\partial x_2} + v_2\frac{\partial u_2}{\partial y_2} + w_2\frac{\partial u_2}{\partial z_2}\right) + \frac{\eta}{\tau}\frac{\partial u_2}{\partial t_2} \qquad (3)$$

For dynamic similitude this must reduce to Eq. (2), with subscripts 2, which means that the dimensionless coefficients of Eq. (3) must divide out. This occurs if

$$\frac{\pi}{\rho'\xi} = \frac{\mu'\eta}{\rho'\xi^2} = \frac{\eta^2}{\xi} = \frac{\eta}{\tau} \qquad (4)$$

The last two terms are obviously the same, since $\eta = \xi/\tau$. Examining the second and third terms,

$$\frac{\mu'}{\rho'}\frac{\eta}{\xi^2} = \frac{\eta^2}{\xi}$$

or

$$\frac{\mu'}{\rho'\eta\xi} = 1$$

Substituting for the dimensionless quantities and rearranging,

$$\frac{\rho_1 u_1 l_1}{\mu_1} = \frac{\rho_2 u_2 l_2}{\mu_2} = R \qquad (5)$$

If the proper base for dynamic pressure be taken, the other relationships given by Eq. (4) can always be satisfied. The dimensionless ratios [Eq. (5)] are given the name *Reynolds number* after Osborne Reynolds, who first determined them. If Reynolds number is the same for two flows having geometrically similar boundaries, the flows are said to be dynamically similar. For example, the flow of the same fluid (same temperature and density) through two pipes, one twice as large as the other, requires that the larger pipe have an average velocity half that of the smaller pipe for dynamic similitude.

The drag on a body can be considered as made up from a pressure difference times an area. Solving Eq. (4) for π,

$$\pi = \frac{\mu'\eta}{\xi} = \rho'\eta^2$$

hence, the drag D can be written in two forms:

$$D = k_1 l^2 \rho u^2 \qquad (6)$$
$$D = k_2 \mu l u \qquad (7)$$

where the ratio of numbers k_1, k_2 is seen to be given by

$$\frac{k_2}{k_1} = \frac{l\rho u}{\mu} = R \qquad (8)$$

for dynamically similar flows. The dimensionless quantities k_1, k_2 are constant for dynamic similarity (one Reynolds number) and, in general, vary with Reynolds number. Rewriting Eqs. (6) and (7),

$$D = f_1(R)\rho l^2 u^2 \qquad (9)$$
$$D = f_2(R)\mu l u \qquad (10)$$

where f_1, f_2 are unknown functions that must, in general, be determined by experiment.

When Reynolds number is very small, the denominator is large compared with the numerator. This means that the viscous terms predominate as compared with the inertial terms. Hence, from Eq. (7), k_2 must not depend on ρ and must therefore be a constant. Similarly, for large Reynolds numbers, the inertial terms predominate, and viscosity should have no effect. Hence, k_1 must be a constant.

These two relations have been observed to hold experimentally. For very large Reynolds numbers, solutions of the Euler equations apply closely, except for a narrow region along the boundaries and possibly in a wake. For very small Reynolds numbers, the inertial terms (those containing density) may be omitted from the Navier-Stokes equations, permitting special solutions to be effected. For the broad range of

Reynolds numbers between the two extremes, theory contributes little and recourse must be had to experimental methods.

116. Flow between Parallel Boundaries. For steady flow between fixed parallel boundaries at low Reynolds numbers, the Navier-Stokes equations can be greatly reduced. Taking the xy-plane midway between the boundaries, the flow normal to the boundaries is everywhere zero; *i.e.*, $w = 0$. Since the velocity at the boundaries must be zero, as shown

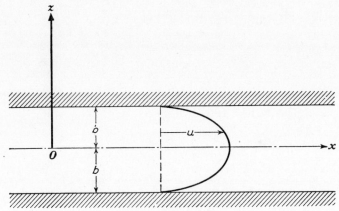

FIG. 129.—Viscous flow between fixed parallel boundaries.

in Fig. 129, the change in velocity with respect to the z-direction is much greater than the change with respect to the xy-directions. Hence, $\dfrac{\partial u}{\partial x}, \dfrac{\partial u}{\partial y}$, $\dfrac{\partial v}{\partial x}, \dfrac{\partial v}{\partial y}$, and their second derivatives are small compared with $\dfrac{\partial u}{\partial z}, \dfrac{\partial v}{\partial z}$, $\dfrac{\partial^2 u}{\partial z^2}, \dfrac{\partial^2 v}{\partial z^2}$ and can be neglected.

Letting the extraneous force potential be $\Omega = gh$, where h is measured vertically upward, then the extraneous forces are given by

$$X = -\frac{\partial}{\partial x}(gh) \ \cdots$$

The incompressible viscous flow equations [Eqs. (25) of the preceding chapter] reduce to

$$\left.\begin{array}{l} \dfrac{\partial}{\partial x}(p + \gamma h) = \mu \dfrac{\partial^2 u}{\partial z^2} \\[2mm] \dfrac{\partial}{\partial y}(p + \gamma h) = \mu \dfrac{\partial^2 v}{\partial z^2} \\[2mm] \dfrac{\partial}{\partial z}(p + \gamma h) = 0 \end{array}\right\} \tag{11}$$

where $\gamma = \rho g$, the unit weight of fluid.

The third equation shows that the pressure varies hydrostatically in the z-direction; *i.e.*, the dynamic pressure does not vary in the z-direction. Hence, in each plane parallel to the xy-plane the pressure gradient does not depend upon z. The flow lines, therefore, have the same direction at corresponding points on any plane parallel to the xy-plane. Since $p + \gamma h$ does not vary with z, the first two equations can be integrated with respect to z as follows:

$$z \frac{\partial}{\partial x} (p + \gamma h) = \mu \frac{\partial u}{\partial z} + C_1$$

$$z \frac{\partial}{\partial y} (p + \gamma h) = \mu \frac{\partial v}{\partial z} + C_2$$

From symmetry, $\frac{\partial u}{\partial z} = 0$, $\frac{\partial v}{\partial z} = 0$, at $z = 0$; hence, $C_1 = C_2 = 0$. Integrating again with respect to z,

$$\frac{z^2}{2} \frac{\partial}{\partial x} (p + \gamma h) = \mu u + C_3$$

$$\frac{z^2}{2} \frac{\partial}{\partial y} (p + \gamma h) = \mu v + C_4$$

Since the flow must be zero at the boundaries,

$$u = v = 0 \qquad \text{for} \qquad z = \pm b$$

where the distance between boundaries is $2b$. Evaluating C_3 and C_4,

$$\left. \begin{aligned} u &= \frac{z^2 - b^2}{2\mu} \frac{\partial}{\partial x} (p + \gamma h) \\ v &= \frac{z^2 - b^2}{2\mu} \frac{\partial}{\partial y} (p + \gamma h) \end{aligned} \right\} \tag{12}$$

the velocity distribution is seen to be parabolic. Rewriting the equations,

$$\left. \begin{aligned} u &= \frac{\partial}{\partial x} \left[(p + \gamma h) \left(\frac{z^2 - b^2}{2\mu} \right) \right] = -\frac{\partial \phi}{\partial x} \\ v &= \frac{\partial}{\partial y} \left[(p + \gamma h) \left(\frac{z^2 - b^2}{2\mu} \right) \right] = -\frac{\partial \phi}{\partial y} \end{aligned} \right\} \tag{13}$$

where $-\phi$ is the quantity in brackets. For this special viscous flow case a velocity potential exists, given by ϕ. Using this as an analogy to potential flow, Hele-Shaw[1] constructed an apparatus consisting of two

[1] H. J. S. Hele-Shaw, Investigation of the Nature of the Surface Resistance of Water and of Stream-line Motion under Certain Experimental Conditions, *Trans. Inst. Naval Architects*, Vol. 40, 1898.

closely spaced glass plates. A transparent fluid is caused to flow between the plates, and dye is continuously injected into the fluid at regular intervals along the upstream edge of the plates. An object placed between the plates causes the fluid to deviate in flowing around it, such that the dyed portions of the fluid trace out streamlines for two-dimensional potential flow. The results are confirmed by potential theory and by other experimental means. There is a slight discrepancy in the immediate neighborhood of the boundaries, which extends outward a distance about equal to the spacing between plates. By decreasing the spacing, the results may be made as accurate as desired.

For motion of the upper plate in the x-direction with velocity U, the boundary conditions become

$$u = v = 0 \quad \text{for} \quad z = -b$$
$$u = U, \quad v = 0 \quad \text{for} \quad z = +b$$

Integrating Eqs. (11) twice and substituting these boundary conditions yield

$$\left. \begin{aligned} u &= \frac{U}{2}\left(1 + \frac{z}{b}\right) + \frac{z^2 - b^2}{2\mu}\frac{\partial}{\partial x}(p + \gamma h) \\ v &= \frac{z^2 - b^2}{2\mu}\frac{\partial}{\partial y}(p + \gamma h) \end{aligned} \right\} \tag{14}$$

The maximum velocity has been displaced from the middle plane. When $p + \gamma h$ is constant, the gradient is zero and flow results due to motion of the upper plate only. The velocity distribution is then linear, and work done in moving the upper plate is converted into heat through viscous shearing of the fluid.

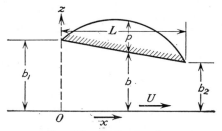

Fig. 130.—Sliding bearing.

117. Theory of Lubrication.

The equations for two-dimensional viscous flow are applicable to the case of a slider bearing and can be applied also to journal bearings. The simple case of a bearing of unit width is developed here, under the assumption that there is no flow out of the sides of the block, *i.e.*, normal to the plane of Fig. 130, where the clearance b is shown to a greatly exaggerated scale. The motion of a bearing block sliding over a plane surface, inclined slightly so that fluid is crowded between the two surfaces, develops large supporting forces normal to the surfaces. The angle of inclination is very small; therefore Eqs. (11) can be applied to give the velocity distribution. Since elevation changes are also very small and flow is in the x-direction only, the equations reduce to

$$\frac{\partial p}{\partial x} = \mu \frac{\partial^2 u}{\partial z^2}, \qquad \frac{\partial p}{\partial z} = 0 \tag{15}$$

It is convenient to consider the inclined block stationary and the plane surface in motion. The two ends of the block (Fig. 130) are considered to be at zero pressure, as a change in static pressure does not affect the computations. Using the notation of Fig. 130, the boundary conditions are

$$x = 0, x = L, p = 0; \qquad u = U, z = 0; \qquad u = 0, z = b$$

Integrating Eqs. (15) twice with respect to z and inserting the boundary conditions to evaluate the constants of integration yield

$$u = \frac{z}{2\mu} \frac{\partial p}{\partial x}(z - b) + U\left(1 - \frac{z}{b}\right) \tag{16}$$

To find the discharge Q, which, by continuity, must be the same at each section, the velocity is integrated over a section of width b,

$$Q = \int_0^b u\,dz = \frac{Ub}{2} - \frac{b^3}{12\mu}\frac{\partial p}{\partial x} \tag{17}$$

for unit width normal to the plane of the figure. At the section having maximum pressure intensity, $\dfrac{\partial p}{\partial x} = 0$ and

$$Q = \frac{Ub_0}{2} \tag{18}$$

where b_0 is the clearance at section of maximum pressure intensity.

Solving Eq. (17) for $\dfrac{\partial p}{\partial x}$ and integrating with respect to x, where $b = b_1 - \alpha x$, $\alpha = (b_1 - b_2)/L$, from Fig. 130,

$$p = \frac{6\mu U}{\alpha(b_1 - \alpha x)} - \frac{6\mu Q}{\alpha(b_1 - \alpha x)^2} + C \tag{19}$$

where Q and C may be determined by the two conditions $p = 0$, for $x = 0, L$. Substituting these conditions,

$$Q = \frac{Ub_1 b_2}{b_1 + b_2} = \frac{Ub_0}{2} \tag{20}$$

and

$$p = \frac{6\mu U x(b - b_2)}{b^2(b_1 + b_2)} \tag{21}$$

The last relation shows that b must be greater than b_2 for positive pressure build-up in the bearing.   From Eq. (20),

$$b_0 = \frac{2b_1 b_2}{b_1 + b_2}$$

$$x\Big]_{p_{max}} = \frac{b_1 - b_0}{\alpha} = \frac{b_1}{\alpha}\frac{b_1 - b_2}{b_1 + b_2} = \frac{b_1 L}{b_1 + b_2}$$

Inserting these values of b_0 and x into Eq. (21) the maximum pressure intensity is

$$p_{max} = \frac{3\mu U L}{2b_1 b_2}\frac{b_1 - b_2}{b_1 + b_2} \tag{22}$$

The force P, which the bearing will sustain, is given by

$$P = \int_0^L p\,dx = \frac{6\mu U}{b_1 + b_2}\int_0^L \frac{x(b - b_2)}{b^2}\,dx$$

Inserting the value of b in terms of x and integrating,

$$P = \frac{6\mu U L^2}{b_2{}^2 (k-1)^2}\left(\ln k - 2\frac{k-1}{k+1}\right) \tag{23}$$

where $k = b_1/b_2$.

The drag D on the bearing is given by

$$D = -\int_0^L \mu \frac{\partial u}{\partial z}\Big|_{z=0}\,dx$$

Evaluating $\dfrac{\partial u}{\partial z}$ for $z = 0$, from Eq. (16),

$$\frac{\partial u}{\partial z}\Big|_{z=0} = -\frac{b}{2\mu}\frac{\partial p}{\partial x} - \frac{U}{b}$$

Inserting the value of the pressure gradient

$$\frac{\partial p}{\partial x} = \frac{6\mu U}{b^2} - \frac{12\mu Q}{b^3}$$

and integrating,

$$D = \frac{\mu U L}{b_2 (k-1)}\left(4\ln k - 6\frac{k-1}{k+1}\right) \tag{24}$$

Equation (23) gives a maximum P for k approximately 2.2.   For this value,

$$P = 0.16\mu \frac{U L^2}{b_2{}^2}, \qquad D = 0.75\mu \frac{U L}{b_2}$$

and the ratio

$$\frac{P}{D} = 0.21 \frac{L}{b_2}$$

can be made very large, since b_2 is small. This type of bearing is capable of sustaining large loads. The pressure distribution is shown for one case in Fig. 130.

For $k = 2.2$, $\bar{x} = 0.58L$, where $\bar{x}$ is the distance to line of action of the bearing load. In general, the position of the line of action is given by

$$\bar{x} = \frac{L}{2}\left[\frac{2k}{k-1} - \frac{k^2 - 1 - 2k \ln k}{(k^2 - 1) \ln k - 2(k-1)^2} \right] \tag{25}$$

Journal bearings[1] are computed in an analogous manner. In general, the clearances are so small compared with the radius of curvature of bearing surface that the equations for plane motion can be applied.

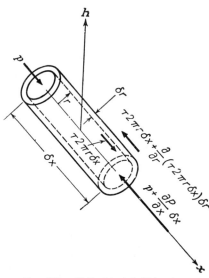

FIG. 131.—Cylindrical fluid lamina.

118. Steady Flow through Circular Tubes. The Navier-Stokes equations may be expressed in cylindrical coordinates and then reduced to the case of one-dimensional flow. It is much simpler, however, to derive the differential equation that applies to this case directly from consideration of a thin cylindrical lamina. For Reynolds numbers less than 2000, based on average velocity and diameter of tube, the flow is laminar and fluid particles move in straight lines parallel to the axis of the tube. For higher Reynolds numbers the flow may be laminar; but in general, it is unstable.

The fluid may be visualized as telescoping layers, with each lamina moving faster than the one next larger. Those fluid particles in contact with the conduit walls are at rest. Selecting the free body as shown in Fig. 131 and writing the equation of motion for the x-direction,

$$2\pi r\ \delta r \left(-\frac{\partial p}{\partial x}\ \delta x \right) + 2\pi r\ \delta r\ \delta x\ \gamma \left(-\frac{\partial h}{\partial x} \right) - \frac{\partial}{\partial r}\left(\tau 2\pi r\ \delta x \right)\delta r = 0$$

[1] For more detailed information see A. E. Norton, "Lubrication," McGraw-Hill Book Company, Inc., New York, 1942.

The first term represents the pressure force on the ends of the element, the second term the action of gravity, and the third term the shear force from the curved surfaces. The resultant of these forces is equated to zero, since each particle moves through the tube with constant speed, *i.e.*, the acceleration is zero. Dividing by the volume of the element,

$$\frac{\partial}{\partial x}\,(p + \gamma h) = -\,\frac{1}{r}\,\frac{\partial}{\partial r}\,(\tau r) \tag{26}$$

The shear stress may be expressed in terms of viscosity and velocity gradient,

$$\tau = -\mu\,\frac{\partial u}{\partial r}$$

where the minus sign is introduced because $\dfrac{\partial u}{\partial r}$ is negative for the particular coordinate system. Substituting into Eq. (26), remembering that $\dfrac{\partial}{\partial x}\,(p + \gamma h)$ is not a function of r and is, in fact, a constant for a given flow, the equation can be integrated twice; thus

$$\frac{r^2}{2}\,\frac{\partial}{\partial x}\,(p + \gamma h) = \mu r\,\frac{\partial u}{\partial r} + C_1$$

and

$$u = \frac{r^2}{4\mu}\,\frac{\partial}{\partial x}\,(p + \gamma h) - C_1\ln r + C_2 \tag{27}$$

Since the velocity must be finite at the axis of the tube, $C_1 = 0$. Solving for C_2, using $u = 0$, for $r = a$,

$$u = -\,\frac{a^2 - r^2}{4\mu}\,\frac{\partial}{\partial x}\,(p + \gamma h) \tag{28}$$

where a is the radius of the tube. The velocity distribution is parabolic, with maximum velocity u_m at the axis:

$$u_m = -\,\frac{a^2}{4\mu}\,\frac{\partial}{\partial x}\,(p + \gamma h) \tag{29}$$

The average velocity $\bar{u}$ can be found by integrating over the section or by recalling that since a paraboloid of revolution has a volume one-half that of its circumscribing cylinder, it must be half the maximum velocity.

$$\bar{u} = -\,\frac{a^2}{8\mu}\,\frac{\partial}{\partial x}\,(p + \gamma h) \tag{30}$$

For a horizontal tube the discharge Q is

$$Q = \frac{\Delta p \pi D^4}{128\mu L} \tag{31}$$

where D is the diameter and Δp the pressure drop in the length L. Equation (31) is usually called *Poiseuille's law;* it was obtained experimentally by Hagen in 1839 and independently by Poiseuille in 1840. The theoretical derivation was made by Wiedemann in 1856.

Equation (27) may be applied to laminar flow in the annular space between two concentric tubes. Letting $r = a$ be the outer boundary and $r = b$ the inner boundary, two conditions are available to determine the two constants:

$$u = 0, \qquad r = a; \qquad u = 0, \qquad r = b$$

Evaluating the constants, the velocity distribution is given by

$$u = -\frac{1}{4\mu}\frac{\partial}{\partial x}(p + \gamma h)\left(a^2 - r^2 + \frac{a^2 - b^2}{\ln b/a}\ln\frac{a}{r}\right) \tag{32}$$

and the discharge Q by

$$Q = \int_b^a 2\pi r \, u \, dr$$
$$= -\frac{\pi}{8\mu}\frac{\partial}{\partial x}(p + \gamma h)\left[a^4 - b^4 - \frac{(a^2 - b^2)^2}{\ln a/b}\right] \tag{33}$$

119. Flow with Very Low Velocity. Percolation. The Navier-Stokes equations for incompressible flow can be solved when the velocity components are considered to be so small that such terms as $\dfrac{u\,\partial u}{\partial x}, \dfrac{v\,\partial u}{\partial y}, \dfrac{w\,\partial u}{\partial z}$ can be neglected. For steady flow, Eqs. (25), Sec. 113, reduce to

$$\left.\begin{array}{l} \dfrac{\partial}{\partial x}(p + \gamma h) = \mu\,\nabla^2 u \\[2mm] \dfrac{\partial}{\partial y}(p + \gamma h) = \mu\,\nabla^2 v \\[2mm] \dfrac{\partial}{\partial z}(p + \gamma h) = \mu\,\nabla^2 w \end{array}\right\} \tag{34}$$

where gravity is the only extraneous force acting. Taking the derivative of the first of the equations with respect to x, the second with respect to y, and the third with respect to z and adding,

$$\nabla^2(p + \gamma h) = \mu\left(\frac{\partial^2}{\partial x^2} + \frac{\partial^2}{\partial y^2} + \frac{\partial^2}{\partial z^2}\right)\left(\frac{\partial u}{\partial x} + \frac{\partial v}{\partial y} + \frac{\partial w}{\partial z}\right)$$

or

$$\nabla^2(p + \gamma h) = \mu\nabla^2(\operatorname{div} q)$$

Since the divergence of the velocity vector is zero in incompressible flow, the equation reduces to

$$\nabla^2(p + \gamma h) = 0 \tag{35}$$

which must be satisfied along with the continuity equation and boundary conditions.   The neglect of the acceleration terms, in effect, states that the complete energy required to overcome viscous shear must be obtained from potential energy.

In the flow of fluid through pervious materials, such as the flow of water through sand, the velocities are usually very small and the passages very small.   Although it is impossible to write the equations for an individual fluid particle, the mass-flow relationship is given by Eq. (35). This is the Laplace equation; hence, a velocity potential exists, with velocity given by

$$\left.\begin{aligned} u &= -\frac{k}{\gamma}\frac{\partial}{\partial x}\,(p + \gamma h) \\[6pt] v &= -\frac{k}{\gamma}\frac{\partial}{\partial y}\,(p + \gamma h) \\[6pt] w &= -\frac{k}{\gamma}\frac{\partial}{\partial z}\,(p + \gamma h) \end{aligned}\right\} \tag{36}$$

where k is the permeability coefficient which has the dimensions of velocity.   The permeability coefficient takes into account the size, shape, spacing, and roughness of the solid particles as well as the physical properties of the fluid.   For two-dimensional flow cases, such as flow under a long dam, the flow net may be constructed graphically or by any convenient means available to ideal fluid theory.   When the porous media is stratified, k takes on directional properties and is no longer a constant.

120. Viscous Flow around a Sphere at Very Low Velocity.   Stokes' Law.   The flow of a viscous fluid around a sphere at very low Reynolds numbers has been solved by Stokes.[1]   The Navier-Stokes equations with the acceleration terms omitted must be satisfied, as well as continuity and the boundary condition that the velocity at the surface of the sphere vanishes.   Equations (34) are the reduced equations that must be satisfied.

The Stokes solution is given by the following equations, which are readily checked by substitution into Eqs. (34), continuity, and the

---

[1] G. Stokes, *Trans. Cambridge Phil. Soc.*, Vol. 8, 1845, and Vol. 9, 1851.

boundary conditions:

$$u = U\left[\frac{3}{4}\frac{ax^2}{r^3}\left(\frac{a^2}{r^2} - 1\right) + 1 - \frac{1}{4}\frac{a}{r}\left(3 + \frac{a^2}{r^2}\right)\right]$$

$$v = U\frac{3}{4}\frac{axy}{r^3}\left(\frac{a^2}{r^2} - 1\right)$$

$$w = U\frac{3}{4}\frac{axz}{r^3}\left(\frac{a^2}{r^2} - 1\right)$$

$$p = -\frac{3}{2}\frac{\mu Uax}{r^3}$$

$$(37)$$

The radius is a; $u = U$ is the undisturbed velocity of the fluid; and p is

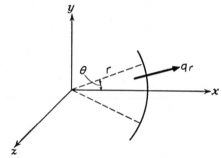

Fig. 132.—Spherical segment for determination of stream function.

the average dynamic pressure intensity at a point. The origin is taken at the center of the sphere.

The Stokes' stream function (Sec. 17) can be obtained from the velocity components. First, the velocity component in the radial direction q_r is given by

$$q_r = u\frac{x}{r} + v\frac{y}{r} + w\frac{z}{r}$$

since x/r, y/r, z/r are direction cosines of a radial line through the point (x,y,z). Substituting for u, v, w from Eqs. (37),

$$q_r = U\left(1 - \frac{3}{2}\frac{a}{r} + \frac{1}{2}\frac{a^3}{r^3}\right)\frac{x}{r} \tag{38}$$

The flow $2\pi\psi$ through the segment of a sphere subtending the angle θ at the origin, as shown in Fig. 132, is given by

$$2\pi\psi = \int_0^\theta 2\pi(-q_r)r\sin\theta\, r\, d\theta$$

Inserting Eq. (38) and performing the integration,

$$\psi = -\frac{U}{2}\left(1 - \frac{3}{2}\frac{a}{r} + \frac{1}{2}\frac{a^3}{r^3}\right)r^2 \sin^2\theta \qquad (39)$$

For translation of the sphere through an infinite fluid, a uniform flow

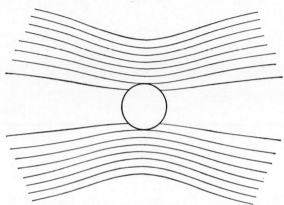

Fig. 133.—Instantaneous streamlines for translation of sphere through viscous fluid.

$u = -U$ may be superposed on the fluid. The stream function for uniform flow is

$$\psi = \frac{Ur^2}{2}\sin^2\theta$$

Adding the two stream functions,

$$\psi = \frac{3}{4}\,Uar\left(1 - \frac{a^2}{3r^2}\right)\sin^2\theta \qquad (40)$$

The instantaneous streamlines are shown in Fig. 133 for equal increments of ψ. These streamlines should be compared with the ideal fluid streamlines shown in Fig. 29. The great difference is not surprising when it is recalled that completely different forces are assumed to act and that the boundary conditions are different.

To find the drag on the sphere the components due to pressure and to shear must be determined and summed up over its surface. Since there is axial symmetry, it is convenient to work with the radial and tangential components of velocity. Referring to Fig. 134 and to Eqs.

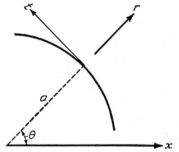

Fig. 134.—Radial and tangential axes on surface of sphere.

(37) and (38),

$$q_r = U \cos \theta \left(1 - \frac{3a}{2r} + \frac{1}{2} \frac{a^3}{r^3} \right) \Bigg\}$$
$$q_t = -U \sin \theta \left[1 - \frac{a}{4r} \left(3 + \frac{a^2}{r^2} \right) \right] \Bigg\} \tag{41}$$

The pressure intensity acting over the surface of the sphere may be found from Eqs. (23), Sec. 112, together with the above equations:

$$\sigma_r = -p + 2\mu \frac{\partial q_r}{\partial r}$$

Evaluating $\frac{\partial q_r}{\partial r}$ at $r = a$ it is found to be zero; hence, the pressure intensity over the surface is given by the expression for average pressure intensity at a point [Eqs. (37)]:

$$p = -\frac{3}{2} \frac{\mu U}{a} \cos \theta \tag{42}$$

Integrating over the surface of the sphere, using the x-component of the pressure force, the drag due to pressure difference is

$$D_p = -2\pi a^2 \int_0^\pi p \cos \theta \sin \theta \, d\theta = 2\pi a \mu U \tag{43}$$

The drag due to viscous shear must be evaluated from the shear stress

$$\tau_{r\theta} = \mu \left(\frac{1}{r} \frac{\partial q_r}{\partial \theta} + \frac{\partial q_t}{\partial r} \right) \tag{44}$$

for $r = a$. Thus

$$\frac{\partial q_r}{\partial \theta} = -U \sin \theta \left(1 - \frac{3a}{2r} + \frac{a^3}{2r^3} \right)$$

which is zero for $r = a$, and

$$\frac{\partial q_t}{\partial r} = -\frac{3}{4} \frac{Ua}{r^2} \sin \theta \left(1 + \frac{a^2}{r^2} \right)$$

which, for $r = a$, becomes

$$\frac{\partial q_t}{\partial r}\bigg|_{r=a} = \frac{-3}{2} \frac{U}{a} \sin \theta$$

The shear stress tangent to the surface and in the plane of symmetry is then

$$\tau_{r\theta} = \frac{-3}{2} \mu \frac{U}{a} \sin \theta \tag{45}$$

The force on the sphere in the x-direction due to shear is given by

$$-2\pi a^2 \int^{\pi} \tau_{r\theta} \sin^2 \theta \, d\theta$$

Substituting for $\tau_{r\theta}|_{r=a}$ and performing the integration give the drag due to viscous shear D_τ:

$$D_\tau = 4\pi a \mu U \tag{46}$$

The total drag on the sphere is then

$$D = 6\pi a \mu U \tag{47}$$

which is known as *Stokes' law*. It should be noted that viscous shear contributes two-thirds of the drag on the sphere.

Using the drag given by Eq. (47), the settling velocity of small particles can be obtained by writing the equation of equilibrium for drag, weight of particle, and buoyant force:

$$6\pi a \mu U + \tfrac{4}{3}\pi a^3 \gamma = \tfrac{4}{3}\pi a^3 \gamma_s$$

where γ_s is the specific weight of the solid particle. Simplifying,

$$U = \frac{2}{9}\frac{a^2}{\mu}(\gamma_s - \gamma) \tag{48}$$

Stokes' law has been found by experiment to hold for Reynolds numbers below 1; *i.e.*,

$$\frac{2a\rho U}{\mu} < 1$$

Exercises

1. Apply the principles of similitude to the Euler equations. What conclusions can be drawn from the equations in this manner?

2. A sleeve 6 in. long is concentric with a 3-in.-diameter shaft and has a clearance of 0.002 in. A fluid with viscosity 1.0 poise completely fills the space between shaft and sleeve.

(a) Find the velocity with which the sleeve moves parallel to the shaft when a force of 20 lb is applied to it parallel to the shaft. The shaft rotates at 1000 rpm relative to the sleeve.

(b) What torque on the shaft is required to overcome resistance between sleeve and shaft?

3. (a) Find the terminal thickness of a liquid film that flows down a glass plate inclined 30 deg with the vertical. The maximum velocity at the surface of the film is 2 ft per sec, $\mu = 0.0001$ lb-sec per ft^2, $\rho = 2.0$ slugs per ft^3.

(b) How many pounds per hour flow down the plate per foot of width?

Ans. (a) 0.032 in.; (b) 826.

4. (a) Find the direction and amount of flow per foot of width between two parallel plates when one is moving relative to the other. The upper plate has a velocity 10 ft per sec in the negative x-direction. The plates are horizontal with a spacing of 0.001 ft. $\frac{dp}{dx} = -10,000$ lb per ft^3; $\mu = 0.0001$ lb-sec per ft^2.

(b) What force is required to move the upper plate per square foot of surface?

5. A vertical turbine shaft carries a load of 80,000 lb on a thrust bearing consisting of 16 flat rocker plates, 3 in. by 9 in., arranged with their long dimension radial from the shaft with their centers on a circle of radius 1.5 ft. The shaft turns at 120 rpm, $\mu = 0.002$ lb-sec per ft^2. If the plates take the optimum angle for maximum load, find (neglecting effects of curvature and radial lubricant flow) (a) the clearance between rocker plate and fixed plate, (b) the maximum pressure intensity under the plates, and (c) the torque loss due to the bearing.

6. Find the drag per foot on the inner tube of a horizontal annular system, where $a = 1.0$ in., $b = 0.5$ in., $\frac{\Delta P}{\Delta L} = -300$ lb per ft^3, $\mu = 0.01$ lb-sec per ft^2, $\rho = 1.8$ slugs per ft^3.

7. A pervious stratum of earth in a valley is 1 mile wide and 40 ft deep. The ground-water table slope is 1 per cent (1 vertical to 100 horizontal). Find the discharge in millions of gallons per day. $k = 50$ ft per day.

8. Plot the same streamlines for steady flow around a sphere, for the two cases: (a) viscous flow, Reynolds number less than 1, and (b) ideal fluid flow, same approach velocity as in (a).

9. What is the largest diameter of dust particle that will settle in standard sea-level air and obey Stokes' law? Find its settling velocity. Specific weight $\gamma = 160$ lb per ft^3.

CHAPTER XII

THE BOUNDARY LAYER

In the preceding chapters two types of fluid problems have been considered: (1) those in which the viscosity has been neglected, *i.e.*, potential or irrotational flow, and (2) those in which viscosity is taken into account but, due to the complexity of the equations, certain inertial terms have been neglected, usually for cases of very low velocities.

In 1904 Prandtl[1] developed the concept of the boundary layer, which provided an important link between the two extreme cases heretofore studied. For fluids having relatively small viscosity, the effect of internal friction in a fluid is appreciable only in a narrow region surrounding the fluid boundaries. From this hypothesis the flow outside the boundary layer could be considered ideal or potential flow. The Navier-Stokes equations apply to the boundary layer and can be reduced to a more simple form. In this chapter the basic equations for the boundary layer are developed and applied to simple flow cases.

121. Description of the Boundary Layer. When motion of a fluid having very small viscosity is started from rest, the flow is essentially irrotational in the first instants. Since the fluid at the boundaries must have zero velocity relative to the boundaries, there is a sharp velocity gradient from the velocity given by the potential flow to the boundary. This velocity gradient in a real fluid sets up shear forces in the boundary that tend to reduce the flow relative to the boundary. That fluid layer that has had its velocity affected by the boundary shear is called the *boundary layer*.

Considering a streamlined body at rest in an otherwise uniform flow, the boundary layer at its upstream end is very thin. As this layer moves along the body, the continual action of the shear forces slows down additional fluid particles, causing an increase in thickness of the boundary layer with distance from the upstream point. The fluid in the layer is subjected to a pressure gradient and to shear forces. The resultant of these forces on a fluid element must equal the time rate of change of its momentum. Basic equations are obtained either from the Navier-Stokes equations or from considerations of momentum.

---

[1] Prandtl, L., Über Flussigkeitsbewegung bei sehr kleiner Reibung, *Verhandl. des* III *Intern. Math.-Kongr.*, Heidelberg, 1904.

In general, the subject of turbulence is beyond the scope of this volume; but in turbulent flow, momentum is brought into the boundary layer by the erratic motion of fluid particles, thereby affecting its growth. This chapter deals with those cases in which the turbulent transfer of momentum can be neglected.

When the body is a thin plate parallel to a uniform stream, the velocity distribution near the plate is shown in Fig. 135. The fluid between the dashed line and the plate has been retarded by shear forces at the boundary surface and comprises the boundary layer. When the motion of fluid particles in this layer are sensibly straight line, it is a laminar boundary layer; and when the motion of fluid particles are erratic, it is a

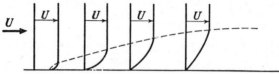

FIG. 135.—Boundary layer growth along a flat plate.

turbulent boundary layer. As the layer grows in thickness, it becomes progressively more unstable and eventually changes from laminar to turbulent. When the boundary layer is turbulent, there is a very thin layer next to the boundary that still has laminar motion. It is called the *laminar sub-layer*.

122. Differential Equation of the Boundary Layer. Restricting the flow to two dimensions, as along a plane boundary, the Navier-Stokes equations [Eqs. (25), Sec. 113] can be greatly simplified. Neglecting extraneous forces and with flow parallel to the xy-plane, the equations become

$$\left.\begin{array}{l} \mu\left(\dfrac{\partial^2 u}{\partial x^2} + \dfrac{\partial^2 u}{\partial y^2}\right) = \dfrac{\partial p}{\partial x} + \rho\left(\dfrac{\partial u}{\partial t} + u\dfrac{\partial u}{\partial x} + v\dfrac{\partial u}{\partial y}\right) \\[2mm] \mu\left(\dfrac{\partial^2 v}{\partial x^2} + \dfrac{\partial^2 v}{\partial y^2}\right) = \dfrac{\partial p}{\partial y} + \rho\left(\dfrac{\partial v}{\partial t} + u\dfrac{\partial v}{\partial x} + v\dfrac{\partial v}{\partial y}\right) \end{array}\right\} \qquad (1)$$

which, together with the continuity equation

$$\frac{\partial u}{\partial x} + \frac{\partial v}{\partial y} = 0 \qquad (2)$$

must be satisfied in the boundary layer. The various terms in the equations will be examined to find their relative importance in the thin boundary layer.

Following Prandtl's[1] method, dimensionless quantities ξ, η, τ of order

[1] W. F. Durand (editor-in-chief), "Aerodynamic Theory," Vol. III, pp. 80–84, Verlag Julius Springer, Berlin, 1935.

of magnitude unity, written 0(1), are introduced. Letting

$$x = l\xi, \qquad y = \delta\eta, \qquad t = \frac{l}{U}\tau$$

where l is characteristic of the length in the direction of flow x; δ is characteristic of the thickness of the boundary layer 0(δ), a distance much smaller than l; and l/U, a time, is characteristic of the flow through the boundary layer. The order of magnitude of v can now be determined relative to u by integration of Eq. (2),

$$v = - \int_0^\delta \frac{\partial u}{\partial x}\, dy = -\frac{\delta}{l} \int_0^1 \frac{\partial u}{\partial \xi}\, d\eta$$

where the dimensionless length parameters have been introduced. The right-hand integral is 0(u), since ξ and η both vary from zero to unity; hence, v is $0\left(\dfrac{\delta}{l}\, u\right)$. Inserting the dimensionless parameters in Eqs. (1),

$$\mu\left(\frac{1}{l^2}\frac{\partial^2 u}{\partial \xi^2} + \frac{1}{\delta^2}\frac{\partial^2 u}{\partial \eta^2}\right) = \frac{1}{l}\frac{\partial p}{\partial \xi} + \rho\left(\frac{U}{l}\frac{\partial u}{\partial \tau} + \frac{u}{l}\frac{\partial u}{\partial \xi} + \frac{v}{\delta}\frac{\partial u}{\partial \eta}\right)$$

$$\mu\left(\frac{1}{l^2}\frac{\partial^2 v}{\partial \xi^2} + \frac{1}{\delta^2}\frac{\partial^2 v}{\partial \eta^2}\right) = \frac{1}{\delta}\frac{\partial p}{\partial \eta} + \rho\left(\frac{U}{l}\frac{\partial v}{\partial \tau} + \frac{u}{l}\frac{\partial v}{\partial \xi} + \frac{v}{\delta}\frac{\partial v}{\partial \eta}\right)$$

Since ξ, η vary from 0 to 1, the second derivatives $\dfrac{\partial^2 u}{\partial \xi^2}$, $\dfrac{\partial^2 u}{\partial \eta^2}$ are 0(u). Therefore, the term $\dfrac{1}{l^2}\dfrac{\partial^2 u}{\partial \xi^2}$ can be neglected in comparison with $\dfrac{1}{\delta^2}\dfrac{\partial^2 u}{\partial \eta^2}$, since δ is much smaller than l. Leaving out of consideration possible shock forces, the terms on the right-hand side may be taken as having the same order of magnitude. The pressure term cannot be of lower order than all the other terms; and since both viscous and inertial terms come into the boundary layer, $\dfrac{\mu}{\delta^2}\dfrac{\partial^2 u}{\partial \eta^2}$ and $\dfrac{\rho u}{l}\dfrac{\partial u}{\partial \xi}$ should be of the same order of magnitude. Furthermore, as $\dfrac{\partial^2 u}{\partial \eta^2}$ and $\dfrac{\partial u}{\partial \xi}$ are 0(u), their coefficients are of the same order; hence,

$$\frac{\mu}{\delta^2} \sim \frac{\rho u}{l}$$

Replacing u by U, the expression may be written

$$\frac{\delta}{l} \sim \sqrt{\frac{\nu}{Ul}} \sim \frac{1}{\sqrt{R}} \tag{3}$$

where $\nu = \mu/\rho$. The ratio of thickness of boundary layer to length of flow is inversely proportional to the square root of Reynolds number, where the characteristic length is measured from the leading edge.

Considering now the second equation, $\frac{1}{\delta}\frac{\partial p}{\partial \eta}$ cannot be of lower order than the acceleration terms, say $\rho\frac{u}{l}\frac{\partial v}{\partial \xi}$. Since v is $0\left(\frac{\delta u}{l}\right)$, $\frac{\partial p}{\partial \eta}$ is

$$0\left(\rho\,\frac{U^2\,\delta^2}{l^2}\right)$$

compared with $\frac{\partial p}{\partial \xi}$ of $0(\rho U^2)$. Hence $\frac{\partial p}{\partial \eta}$ is very small compared with $\frac{\partial p}{\partial \xi}$ and can be set equal to zero; or in words, the pressure intensity does not vary with y and is determined by the potential flow outside the boundary layer.

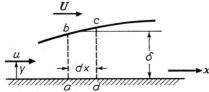

Fig. 136.—Segment of boundary layer.

The final equations may now be written, *viz.*,

$$\mu\,\frac{\partial^2 u}{\partial y^2} = \frac{\partial p}{\partial x} + \rho\left(u\,\frac{\partial u}{\partial x} + v\,\frac{\partial u}{\partial y} + \frac{\partial u}{\partial t}\right) \tag{4}$$

and

$$\frac{\partial u}{\partial x} + \frac{\partial v}{\partial y} = 0 \tag{5}$$

By introducing the stream function ψ,

$$u = -\,\frac{\partial \psi}{\partial y}, \qquad v = \frac{\partial \psi}{\partial x}$$

which identically satisfies the continuity equation, Eq. (4) becomes

$$-\mu\,\frac{\partial^3 \psi}{\partial y^3} = \frac{\partial p}{\partial x} + \rho\left(\frac{\partial \psi}{\partial y}\frac{\partial^2 \psi}{\partial x\,\partial y} - \frac{\partial \psi}{\partial x}\frac{\partial^2 \psi}{\partial y^2} - \frac{\partial^2 \psi}{\partial y\,\partial t}\right) \tag{6}$$

It must be remembered that this equation holds only for a thin boundary layer, which from Eq. (3) occurs for large Reynolds numbers.

123. Momentum Equation Applied to the Boundary Layer. Following von Kármán's[1] method, the principle of momentum may be applied directly to the boundary layer, without recourse to the Navier-Stokes equations. Considering a small segment of the layer (Fig. 136), where

[1] Th. von Kármán, On Laminar and Turbulent Friction, *Z. angew. Math Mech.*, Bd. 1, pp. 235–236, 1921.

$abcd$ is fixed, the resultant force in the x-direction must equal the time rate of increase of momentum within $abcd$ minus the net influx of momentum across the surface of the element in unit time. The resultant force on the element is, for unit breadth,

$$-\mu \left.\frac{\partial u}{\partial y}\right|_{y=0} dx - \frac{\partial p}{\partial x} dx \; \delta$$

The net mass outflow through cd and ab is

$$\frac{\partial}{\partial x} \left(\int_0^\delta \rho u \, dy \right) dx$$

This mass must be entering through bc and hence brings into the element in unit time the momentum

$$U \frac{\partial}{\partial x} \left(\int_0^\delta \rho u \, dy \right) dx$$

The excess of momentum per unit time leaving cd over that entering ab is

$$\frac{\partial}{\partial x} \left(\int_0^\delta \rho u^2 \, dy \right) dx$$

The time rate of increase of momentum within $abcd$ is given by

$$\left(\int_0^\delta \rho \frac{\partial u}{\partial t} \, dy \right) dx$$

Assembling the force and momentum terms, then dividing out dx,

$$\int_0^\delta \rho \frac{\partial u}{\partial t} \, dy + \frac{\partial}{\partial x} \int_0^\delta \rho u^2 \, dy - U \frac{\partial}{\partial x} \int_0^\delta \rho u \, dy = -\frac{\partial p}{\partial x} \delta - \mu \left.\frac{\partial u}{\partial y}\right|_{y=0} \quad (7)$$

which is the desired momentum equation.

Equation (7) can also be obtained directly from Eq. (4) through integration with respect to y. First,

$$\int_0^\delta \left(u \frac{\partial u}{\partial x} + v \frac{\partial u}{\partial y} \right) dy = \int_0^\delta u \frac{\partial u}{\partial x} \, dy + uv \Big]_0^\delta - \int_0^\delta u \frac{\partial v}{\partial y} \, dy$$

where the second integration was effected by parts. Since $u = v = 0$, for $y = 0$, and

$$\left. v \right|_{y=\delta} = \int_0^\delta \frac{\partial v}{\partial y} \, dy = - \int_0^\delta \frac{\partial u}{\partial x} \, dy$$

from Eq. (5),

$$uv \Big]_0^\delta = -U \int_0^\delta \frac{\partial u}{\partial x} \, dy$$

Also, using Eq. (5),

$$-\int_0^\delta u\frac{\partial v}{\partial y}\,dy = \int_0^\delta u\frac{\partial u}{\partial x}\,dy = \frac{1}{2}\int_0^\delta \frac{\partial u^2}{\partial x}\,dy$$

Equation (7) is now obtained by multiplying Eq. (4) by dy and integrating, using these relationships.

For steady flow of an infinite fluid along a flat plate the pressure drop is zero, U becomes a constant, and Eq. (7) reduces to

$$\tau_0 = \frac{\partial}{\partial x}\int_0^\delta \rho(U - u)u\,dy \tag{8}$$

where

$$\tau_0 = \mu\frac{\partial u}{\partial y}\bigg]_{y=0}$$

The momentum approach is highly desirable when the equations of motion are too difficult to handle or when little is known of the internal mechanism of the phenomenon. Equation (8) is applied to flow along a flat plate in Sec. 128.

124. Diffusion of Vorticity from a Boundary. The concept of the development and growth of the boundary layer can be clarified by an analogy between diffusion of heat and diffusion of vorticity in a fluid stream.

Potential flow may be shown to satisfy the Navier-Stokes equations. Introducing the velocity potential into the terms containing viscosity,

$$\nu\nabla^2 u = -\nu\frac{\partial}{\partial x}\nabla^2\phi \cdots$$

Since $\nabla^2\phi = 0$ for potential flow, the terms containing viscosity drop out, leaving the Euler equations.

The difficulty, however, is with the boundary conditions. A viscous fluid clings to the boundary and requires two conditions, that the normal and tangential components of velocity at the boundary vanish relative to the boundary. These conditions can in general be met by the Navier-Stokes equations, as they are of higher order than the Euler equations. With the assumption of potential flow, however, they cannot be satisfied, since their order is reduced. In fact, it has been shown by Jeffreys[1] that when a velocity potential is assumed, with no motion of fluid at the boundaries relative to the boundaries, the only motion is one of translation of fluid and boundary as if both were solid.

In the first instants as motion is started, the flow is irrotational. Due to the lack of slip at the boundaries, rotation starts there. For a plane

[1] Jeffreys, *Proc. Roy. Soc.* A., Vol. 128, p. 376. 1930.

boundary the rotation ω_z is given by

$$\omega_z = \frac{1}{2}\left(\frac{\partial v}{\partial x} - \frac{\partial u}{\partial y}\right) \tag{9}$$

Differentiating the first of the two-dimensional Navier-Stokes equations [Eqs. (1)] by y and the second by x, then subtracting the first from the second, using Eqs. (2) and (9),

$$u\frac{\partial \omega_z}{\partial x} + v\frac{\partial \omega_z}{\partial y} + \frac{\partial \omega_z}{\partial t} = \nu\left(\frac{\partial^2 \omega_z}{\partial x^2} + \frac{\partial^2 \omega_z}{\partial y^2}\right) \tag{10}$$

Replacing ω_z by a symbol for temperature and ν by a symbol for thermal diffusivity, the equation gives the variation in temperature of a fluid with velocity components u,v. Therefore, heat is imparted to a fluid from a body in the flow in exactly the same manner that vortices diffuse into the fluid. For very slow motions, *i.e.*, low Reynolds numbers, heat flows out in all directions from the boundary and, similarly, vortices spread in all directions from the boundary, making the flow rotational. For high velocities or very small viscosities, *i.e.*, high Reynolds numbers, the only fluid heated would be in the narrow layer of fluid surrounding the body and in the wake. Similarly, rotational flow is confined to the narrow layer called the boundary layer and to the wake.

Although this analogy does not aid in the quantitative solution of the Navier-Stokes equations, it presents a clear picture of the spread of vorticity.

Referring to Eq. (9), $\frac{\partial v}{\partial x}$ is small compared with $\frac{\partial u}{\partial y}$ in the boundary layer. Hence, the rotation term becomes

$$\omega_z = -\frac{1}{2}\frac{\partial u}{\partial y} \tag{11}$$

which is large in a thin boundary layer. At the outer edge of the boundary layer, as u approaches the velocity U of the main flow, $\frac{\partial u}{\partial y}$ becomes small and the vorticity vanishes.

125. Definition of Boundary-layer Thickness. Although the velocity u in the boundary layer approaches the velocity U of the main flow asymptotically, for practical purposes the boundary layer may be defined as the region where most of the velocity change occurs. Various definitions of boundary-layer thickness have been suggested.

The most basic definition is referred to the displacement of the main flow due to slowing down of fluid particles in the boundary zone. This thickness δ_1 is given by

$$U\delta_1 = \int (U - u)\, dy \qquad (12)$$

Referring to Fig. 137a, the line $y = \delta_1$ is drawn such that the shaded areas are equal. This distance is, in itself, not the distance that is strongly affected by the boundary. In fact, that region is frequently taken as $3\delta_1$.

Another definition, given by Fig. 137b, is the distance to the point where $u/U = 0.99$. A third definition, given by Fig. 137c, is the dis-

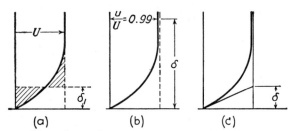

Fig. 137.—Definitions of boundary layer thickness.

tance from the wall to the intersection of the tangent to the velocity distribution curve at the origin and the asymptote.

126. Application of Equations to Curved Boundaries. The differential equations, worked out for a thin boundary layer, are equally applicable to curved boundaries when the radius of curvature is large compared with the boundary-layer thickness. Centrifugal forces enter into the expression for pressure change normal to the boundary, but the actual change is $0(\delta)$. The coordinate x may be taken as the distance along the curved surface, and y as the normal distance. For curved boundaries, U and $\dfrac{\partial p}{\partial x}$, in general, become functions of x.

127. Two-dimensional Flow along a Flat Plate: Exact Solution. For steady flow of an infinite fluid along a flat plate, the pressure gradient becomes zero and the boundary-layer equations become integrable. Equations (4) and (5) become

$$u\,\frac{\partial u}{\partial x} + v\,\frac{\partial u}{\partial y} = \nu\,\frac{\partial^2 u}{\partial y^2} \qquad (13)$$

and

$$\frac{\partial u}{\partial x} + \frac{\partial v}{\partial y} = 0 \qquad (14)$$

The undisturbed velocity U (Fig. 138) is constant.   To satisfy continuity, the stream function is introduced:

$$u = -\frac{\partial\psi}{\partial y}, \qquad v = \frac{\partial\psi}{\partial x}$$

The assumption is next made that the velocity distribution curves in the boundary layer all have the same form; *viz.*,

$$\frac{u}{U} = F\left(\frac{y}{\delta}\right) = F(\eta)$$

where $\eta = y/\delta$, δ is the thickness of boundary layer, and F is an unknown

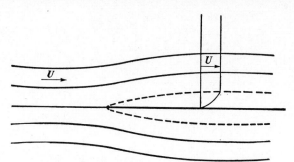

Fig. 138.—Flow of an infinite fluid along a flat plate.

function.   In Sec. 122 it was shown that

$$\frac{\delta}{x} \sim \frac{1}{\sqrt{R}}, \qquad R = \frac{Ux}{\nu}$$

hence,

$$\frac{u}{U} = F\left(\frac{y}{x}\sqrt{R}\right)$$

where δ has absorbed the proportionality factor.   For convenience it is now assumed that $F(\eta)$ is the derivative of another function $f(\eta)$,

$$\frac{u}{U} = f'(\eta), \qquad \eta = \frac{y}{x}\sqrt{R} \qquad (15)$$

The stream function is obtained by integration of Eq. (15), as follows:

$$u = Uf'(\eta) = -\frac{\partial\psi}{\partial y}$$

then

$$\psi = -U\!\int\! f'(\eta)\,dy + G(x) \qquad (16)$$

where $G(x)$ is an arbitrary function of x only and may be taken as zero, since it is convenient to let the boundary streamline be $\psi = 0$. From the value of η, Eq. (15)

$$d\eta = dy \, \frac{\sqrt{R}}{x}$$

Substituting for dy in Eq. (16), integrating, and simplifying,

$$\psi = -\sqrt{\nu U x}\, f(\eta) \tag{17}$$

where, at the boundary, $f(\eta) = f(0) = 0$

The terms $\dfrac{\partial u}{\partial x}$, $\dfrac{\partial u}{\partial y}$, $\dfrac{\partial^2 u}{\partial y^2}$, and v are determined from ψ for substitution into Eq. (13), as follows:

$$\frac{\partial u}{\partial x} = U f''(\eta) \frac{\partial \eta}{\partial x} = -\frac{1}{2} \frac{U^{3/2} y f''(\eta)}{\sqrt{\nu}\, x^{3/2}}$$

$$f''(\eta) = \frac{\partial^2 f(\eta)}{\partial \eta^2}$$

$$\frac{\partial u}{\partial y} = U f''(\eta) \frac{\partial \eta}{\partial y} = \frac{U \sqrt{R}}{x} f''(\eta)$$

and

$$\frac{\partial^2 u}{\partial y^2} = \frac{UR}{x^2} f'''(\eta), \qquad f'''(\eta) = \frac{\partial^3 f(\eta)}{\partial \eta^3}$$

Finally,

$$v = \frac{\partial \psi}{\partial x} = -\frac{1}{2} \sqrt{\frac{\nu U}{x}} f(\eta) + \frac{y}{2x} U f'(\eta)$$

Substitution of these terms in Eq. (13) yields, upon simplification,

$$f(\eta) f''(\eta) + 2 f'''(\eta) = 0 \tag{18}$$

This differential equation of the velocity distribution in the boundary layer is of the third order; and hence, its solution has three constants of integration. The boundary values are $u = v = 0$, for $y = 0$, and $u = U$, for $y \to \infty$, which become, in terms of η,

$$f'(\eta) = 1, \quad \text{for} \quad \eta \to \infty$$
$$f(\eta) = f'(\eta) = 0 \quad \text{for} \quad \eta = 0$$

Equation (18) has been solved by Blasius[1] by building up particular solutions that satisfy the boundary conditions. The equation was integrated numerically by Töpfer,[2] using the boundary conditions at $\eta = 0$.

---

[1] H. Blasius, Grenzschichten in Flüssigkeiten mit kleiner Reibung, *Z. Math. Physik*, Bd. 56, pp. 4–13, 1908.

[2] C. Töpfer, *Z. Math. Physik*, Bd. 60, p. 397, 1912.

The results are practically identical in both cases, but the methods of solution are so involved that reference is made to the original papers for details.   The numerical results are given in the following table and have been found to agree closely with experiment.

THEORETICAL VELOCITY DISTRIBUTION IN BOUNDARY LAYER

$$\eta = \frac{y}{x}\sqrt{R} \qquad \frac{u}{U} = f'(\eta)$$

η	$f'(\eta)$	η	$f'(\eta)$	η	$f'(\eta)$	η	$f'(\eta)$
0	0	1.6	0.5168	3.2	0.8761	4.8	0.9878
0.2	0.0664	1.8	0.5778	3.4	0.9018	5.0	0.9919
0.4	0.1328	2.0	0.6298	3.6	0.9233	5.2	0.9943
0.6	0.1990	2.2	0.6813	3.8	0.9411	5.4	0.9962
0.8	0.2647	2.4	0.7290	4.0	0.9555	5.6	0.9975
1.0	0.3298	2.6	0.7725	4.2	0.9670	5.8	0.9984
1.2	0.3938	2.8	0.8115	4.4	0.9759	6.0	0.9990
1.4	0.4563	3.0	0.8461	4.6	0.9827		

Blasius has computed the displacement of the main flow [Eq. (12)] to be

$$\delta_1 = 1.73\sqrt{\frac{\nu x}{U}}$$

or

$$\frac{\delta_1}{x} = \frac{1.73}{\sqrt{R}} \tag{19}$$

Letting the thickness of the boundary layer be taken arbitrarily as $3\delta_1$, then

$$\frac{\delta}{x} = \frac{3\delta_1}{x} = \frac{5.2}{\sqrt{R}} \tag{20}$$

which corresponds to a value $u/U = 0.994$ from the theoretical solution.

From the table, the value of $\dfrac{\partial u}{\partial y}$ can be evaluated for $y = 0$; thus

$$\frac{\Delta u}{\Delta y} = \frac{U\Delta f'(\eta)}{\dfrac{x}{\sqrt{R}}\Delta(\eta)}$$

Since the function is a straight line for small values of η, the increment $\Delta\eta = 0.2$ may be taken

$$\left.\frac{\partial u}{\partial y}\right|_{y=0} = \frac{0.0664}{0.20}\frac{U\sqrt{R}}{x}$$

and the shear stress at the boundary τ_0 becomes

$$\tau_0 = \mu \frac{\partial u}{\partial y}\bigg]_{y=0} = 0.332 \sqrt{\frac{\rho\mu U^3}{x}} \tag{21}$$

The shear stress at the boundary is observed to decrease with distance from the leading edge, while the boundary layer thickness increases.

The drag on one side of a flat plate per unit breadth is

$$D = \int_0^l \tau_0 \, dx = 0.332 \sqrt{\rho\mu U^3} \int_0^l \frac{dx}{\sqrt{x}}$$

hence,

$$D = 0.664 \sqrt{\rho\mu U^3 l} \tag{22}$$

or

$$D = \frac{0.664}{\sqrt{R}} \rho U^2 l \tag{23}$$

Expressing the drag in terms of a drag coefficient C_D times the stagnation pressure $\rho U^2/2$ and the area of plate l (per unit breadth),

$$D = C_D \frac{\rho U^2}{2} l$$

where

$$C_D = \frac{1.328}{\sqrt{R}} \tag{24}$$

128. Momentum Equation Applied to Two-dimensional Flow along a Flat Plate. Rewriting Eq. (8), with the limit of integration 0 to h,

$$\tau_0 = \frac{\partial}{\partial x} \int_0^h \rho(U - u) u \, dy \tag{25}$$

where h is greater than δ but is independent of x. The quantity h is still of the order of magnitude of δ, but its exact value is unimportant, since the integrand vanishes as u approaches U.

The momentum equation gives no information regarding the velocity distribution in the boundary layer. For an assumed distribution, which satisfies the boundary conditions $u = 0$, $y = 0$, $u = U$, $y = \delta$, the boundary layer thickness as well as the shear at the boundary can be determined.

Letting

$$\frac{u}{U} = F\left(\frac{y}{\delta}\right) = F(\eta), \qquad \eta = \frac{y}{\delta} \tag{26}$$

as before, where δ is unknown, the integral of Eq. (25) can be calculated:

$$\int_0^h \rho(U - u)\, u \, dy = \delta U^2 \rho \int_0^{h/\delta} \left(1 - \frac{u}{U}\right) \frac{u}{U} \, d\frac{y}{\delta}$$

$$= \delta U^2 \rho \int_0^{h/\delta} (1 - F)F \, d\eta$$

Once the function F is assumed, the integral becomes a constant whose value will be given the symbol α. Then

$$\tau_0 = \frac{\partial}{\partial x} (\delta U^2 \rho \alpha)$$

Now

$$\tau_0 = \mu \frac{\partial u}{\partial y}\bigg]_{y=0} = \frac{\mu U}{\delta} F'(0)$$

from Eq. (26). Letting $F'(0)$ be represented by β and equating the expressions for τ_0,

$$\frac{\partial}{\partial x} (\delta U^2 \rho \alpha) = \frac{\mu U \beta}{\delta}$$

Simplifying,

$$\delta \frac{d\delta}{dx} = \frac{\nu \beta}{\alpha U}$$

and performing the integration for $\delta = 0$, $x = 0$,

$$\delta = \sqrt{\frac{2\nu\beta x}{\alpha U}} \tag{27}$$

The surface shear stress now becomes

$$\tau_0 = \sqrt{\frac{\alpha\beta}{2} \frac{\mu\rho U^3}{x}} \tag{28}$$

and the drag on one side of the plate, per unit breadth,

$$\int_0^l \tau_0 \, dx = \sqrt{2\alpha\beta\mu\rho U^3 l} \tag{29}$$

The values of δ and τ_0 are evaluated for two velocity distributions. First, following Lamb,

$$F(\eta) = \sin \frac{\pi}{2} \eta, \qquad 0 \leqslant \eta \leqslant 1$$

$$F(\eta) = 1, \qquad\quad 1 < \eta < \frac{h}{\delta}$$

Then

$$\alpha = \int_0^1 \left(1 - \sin \frac{\pi}{2} \eta\right) \sin \frac{\pi}{2} \eta \, d\eta = 0.137$$

and

$$\beta = F'(0) = \frac{\pi}{2} \cos \frac{\pi}{2} \eta \Big|_{\eta=0} = \frac{\pi}{2}$$

Hence

$$\delta = \frac{4.8x}{\sqrt{R}}, \qquad \tau_0 = 0.327 \sqrt{\frac{\mu \rho U^3}{x}}$$

The displacement of streamlines is, from Eq. (12),

$$\delta_1 = \int_0^\delta \left(1 - \frac{u}{U}\right) dy = \delta \int_0^1 \left(1 - \frac{u}{U}\right) d\left(\frac{y}{\delta}\right)$$

$$= \delta \int_0^1 (1 - F) \, d\eta \qquad\qquad (30)$$

which gives

$$\delta_1 = 0.365\delta = 1.752 \frac{x}{\sqrt{R}}$$

Another velocity distribution, suggested by Prandtl, is

$$F = \frac{3}{2} \eta - \frac{\eta^3}{2}, \qquad 0 \leq \eta \leq 1, \qquad F' = \frac{3}{2} - \frac{3}{2} \eta^2$$

$$F = 1, \qquad\qquad 1 < \eta, \qquad F'(0) = \tfrac{3}{2} = \beta$$

$$\alpha = \int_0^1 \left(1 - \frac{3}{2}\eta + \frac{\eta^3}{2}\right)\left(\frac{3}{2}\eta - \frac{\eta^3}{2}\right) d\eta = 0.139$$

Hence,

$$\delta = \frac{4.65x}{\sqrt{R}}, \qquad \tau_0 = 0.322 \sqrt{\frac{\mu \rho U^3}{x}}$$

and

$$\delta_1 = 0.375\delta = 1.743 \frac{x}{\sqrt{R}}$$

Comparing these values with the exact values [Eqs. (19) and (20)], the agreement is remarkable. The results are comparatively insensible to the particular velocity distribution selected.

The momentum method, due to its extreme simplicity, becomes a very useful tool, particularly for those cases where the exact method cannot be evaluated.

129. Boundary-layer Growth with Pressure Rise. Separation. When the pressure intensity decreases in the direction of motion, the pressure force tends to accelerate the boundary layer and thereby aids in keeping it thin.   With an adverse pressure gradient the pressure and boundary shear both act to retard the layer, thereby causing it to thicken. If there is not sufficient kinetic energy available in the layer to carry it through the zone of increasing pressure, the fluid particles near the wall lose their forward velocity, *i.e.*, $\dfrac{\partial u}{\partial y}\bigg]_{y=0} = 0$, and the streamline moving along near the wall is diverted out from the boundary.   This phenomenon is known as *separation*.   When separation occurs, the boundary layer becomes thick, and vorticity is shed from the boundary into the thick-

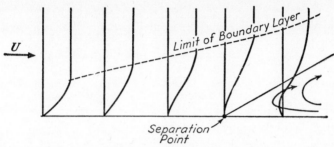

Fig. 139.—Effect of adverse pressure gradient on boundary layer.   Separation.

ened boundary layer.   Back flow necessarily occurs downstream from the point of separation, and eddies develop in the boundary layer.

The action of an adverse pressure gradient on the boundary layer is illustrated in Fig. 139.   The slope of the velocity distribution curve grows more steep until it becomes vertical, at which point separation occurs.   There the streamline leaves the boundary, and back flow commences.

Limiting the discussion to steady flow, the differential equation for the boundary layer [Eq. (4)] can be applied to the boundary, *i.e.*, let $u = v = 0$; hence

$$\frac{dp}{dx} = \mu\,\frac{\partial^2 u}{\partial y^2} \tag{31}$$

Furthermore, the pressure gradient may be expressed in terms of the potential flow outside the boundary layer, using Bernoulli's equation:

$$p + \rho\,\frac{U^2}{2} = C, \qquad \frac{\partial p}{\partial x} = -\rho U\,\frac{\partial U}{\partial x} \tag{32}$$

Equation (31) now takes the form

$$\nu\,\frac{\partial^2 u}{\partial y^2} = -U\,\frac{\partial U}{\partial x} \tag{33}$$

which applies only to the boundary; *i.e.*, $y = 0$.

Substituting the expression for pressure gradient [Eq. (32)] into the momentum equation [Eq. (7)] and solving for $\tau_0 = \mu \dfrac{\partial u}{\partial y}\Big]_{y=0}$,

$$\tau_0 = \rho U \frac{\partial U}{\partial x}\, \delta - \frac{\partial}{\partial x}\int_0^\delta \rho u^2\, dy + U \frac{\partial}{\partial x}\int_0^\delta \rho u\, dy \qquad (34)$$

Equations (33) and (34) together with Eqs. (4) and (5) are the controlling equations for determining boundary layer growth when the velocity U in the main flow is a known function of x.

When the velocity outside the boundary layer is increasing in the direction of flow, $\dfrac{\partial^2 u}{\partial y^2}$ is negative, from Eq. (33), which means that the slope of the velocity distribution curve at the boundary is decreasing; hence, the trend in the boundary layer is opposite that shown in Fig. 139, and separation cannot occur. With a pressure rise in the direction of flow, $\dfrac{\partial^2 u}{\partial y^2}$, from Eq. (31), is positive, and the sequence of events illustrated in Fig. 139 is sure to occur, provided that the adverse pressure zone exists over a sufficiently long portion of the boundary.

A quantitative calculation can be carried out for each particular case to the point of separation. This is effected by expressing U as a power series in x, by introducing the stream function, and by substituting in Eq. (4). The methods are involved, and reference is made to the work of Blasius[1] for details. It should be remarked that the methods do not give information beyond the point of separation, since Eq. (4) is valid only for thin boundary layers. Hiemenz[2] has carried through the theoretical calculations and has also conducted very careful experiments on the flow of water around a circular cylinder. He found that his calculations located the point of separation within one degree of that deter‧mined by experiment.

Beyond the separation point the skin friction forces are generally negligible, and the fluid may be again considered as frictionless, but with vorticity in the wake.

Equations (33) and (34) have been used by Pohlhausen[3] with a velocity distribution of the form

$$u = ay + by^2 + cy^3 + dy^4$$

[1] H. Blasius, Grenzschichten in Flüssigkeiten mit kleiner Reibung, *Z. Math. Physik*, Bd. 56, pp. 4–13, 1908.

[2] Hiemenz, Die Grenzschicht an einem in den gleichförmigen Flüssigkeitsstrom eingetauchten geraden Kreiszylinder, *Dinglers polytech. J.*, Vol. 326, p. 321, 1911.

[3] Pohlhausen, K., Zur näherungsweisen Integration der Differentialgleichung der laminaren Grenzschicht, *Z. angew. Math. Mech.*, Bd. 1, p. 252, 1921.

The calculations are quite involved but agreed with the results of Hiemenz on location of separation point within 1 per cent.

Further calculations by Prandtl[1] show that the boundary layer is incapable of sustaining itself without separation against strong adverse pressures. For most technical applications where it is desirable to convert kinetic energy into pressure energy by means of an expanding conduit, the action of "turbulence" increases the momentum carried into the boundary layer from the main flow and, hence, reduces the action of adverse pressure gradient in causing separation.

130. Turbulence. The resulting flows obtained by use of the Navier-Stokes equations are not necessarily stable and, in fact, for large Reynolds numbers do not occur. This does not mean that the Navier-Stokes equations are inadequate to describe the flow phenomena but that the method of applying them, *i.e.*, by considering the flow is steady, is inaccurate. For large Reynolds numbers the smooth streaming flow considered to take place with viscous fluids is broken down into a flow in which the paths of individual particles are erratic, and there are rapid fluctuations of velocity and pressure at any point.

If flow is considered to start at low Reynolds numbers and then increase, turbulence develops from the shedding of vortices from the boundaries. At high Reynolds numbers, due to the complexity of vortices in the flow and their interactions and decay, an apparently erratic condition is produced, although the temporal mean velocity and pressure may be constant.

The particular value of Reynolds number at which turbulence causes a change in the flow relationships depends on the type of flow considered and upon the characteristic length and velocity selected for Reynolds number. For flow through straight round tubes the motion is laminar for $VD/\nu < 2000$, where V is the average velocity and D the diameter. It may be laminar for higher values of Reynolds numbers, but it is then unstable and breaks down into turbulent flow when disturbed, as by a change in flow direction.

For the boundary layer, it is generally laminar for values of Ux/ν up to about 300,000 to 500,000, where x is measured from the upstream point of the boundary.

The erratic motion of the fluid particles in turbulent flow transfers momentum through the fluid and thereby creates apparent shear stresses that may be much greater than those due to viscosity alone. Turbulence acts in such a manner that momentum is carried from the central portions of the fluid toward the boundaries, causing a greater velocity gradient at

---

[1] W. F. Durand (editor-in-chief), "Aerodynamic Theory," Vol. III, pp. 112–119, Verlag Julius Springer, Berlin, 1935.

the boundaries, which in turn causes larger boundary drag.   Its action in carrying momentum into the boundary layer also delays separation and may decrease the size of the wake.   In fact, the total drag on a body may be less when a turbulent boundary layer develops.   A good example is the flow of an infinite fluid around a sphere.   The total drag on the sphere reduces when the laminar boundary layer becomes turbulent and moves the separation point downstream, thereby decreasing the size of the wake.

INDEX

A

Acceleration, 3
Acceleration components, 11–14, 29
Adverse pressure gradient, 254–257
"Aerodynamic Theory," 242n.
Airfoil, Joukowski, 150–155
  extended, 154–155
Analytical statics of three-dimensional continuum 208–215
Angular deformation, 1, 217–218
Annular space, flow through, 234
Argand diagram, 85
Argument of complex number, 86

B

Back flow, 255
Bearing, journal, 229, 232
  sliding, 229
Bernoulli equation, 23–26, 28, 181
Blasius, H., 250n., 251, 256n.
Blasius theorem, derivation of, 137–141
  example of, 143
Borda's mouthpiece, 169–174
Boundary, diffusion of vorticity from, 246–247
Boundary conditions, 10
  dynamical, 19, 28, 223
  kinematical, 16–19, 30, 208, 223
  for rotation of cylinder, 132–133
  for translation of cylinder, 125–126
Boundary layer, 29, 208, 241–258
  curved, 248
  definition of thickness of, 247–248
  description of, 241–242
  differential equation of, 242
  growth of, 254–257
  laminar, 242
  laminar sub-layer, 242
  momentum equation applied to, 244–246
  turbulent, 242

Boundary layer, velocity distribution in, 251, 253–254
Bounding streamline, 166

C

Cauchy integral theorem, 47, 139–140
Cauchy-Riemann equations, 89–91, 99–100, 139
Channel, flow into, with diverging walls, 122–123
  with rectangular walls, 121–122
Circular arc, steady flow around, 144–150
Circular cylinder, separation point of, 256
  steady flow around, 141–144
Circular tube, steady flow through, 232–234
Circular vortex ring, 195–196
Circulation, 47, 49–51, 142–144
  elliptic, 120–121
  about vortex tube, 183–184
Coefficient, of contraction, 174, 176
  permeability, 235
Cohen, A., 41n.
Complex potential, 96
Complex variables, 85–93
Complex velocity, 96–100
Conformal mapping, 93–100
Conjugate complex numbers, 89
Conjugate functions, 92
Connected region, multiply, 140n.
  simply, 126n.
Continuity equation, 10, 14–16, 36
Continuous function, 8
Continuum, 3
  analytical statics of, 208–215
Contraction, coefficient of, 174, 176
Coordinates, elliptic, 119–121, 131
  plane polar, 51
  scale factors for two-dimensional, 129–131
  spherical polar, 40–41, 62
Curl, of velocity vector, 20

259

Joseph S. Mai

July 16, 196?

New York